Self Reliance Books

Get more historic titles on animal and stock breeding, gardening and old fashioned skills by visiting us at:

http://selfreliancebooks.blogspot.com/

Introduction

I am pleased to present another title in the "Raising Pigs" series..

As with all reprinted books of this age that are intended to perfectly reproduce the original edition, considerable pains and effort had to be undertaken to correct fading and sometimes outright damage to existing proofs of this title. At times, this task is quite monumental, requiring an almost total "rebuilding" of some pages from digital proofs of multiple copies. Despite this, imperfections still sometimes exist in the final proof and may detract from the visual appearance of the text.

I hope you enjoy reading this book as much as I enjoyed re-publishing and making it available to fanciers again.

With Regards,

Jackson Chambers

CONTENTS.

INTRODUCTION . . . . Page. vii

PART I.

CHAP. I. *Of the Natural History of the Hog* . 1

CHAP. II. *The Rise and Progress of Manufacturing Pork into Flitches and Hams in Dumfriesshire, and sending it to the London Market, with a Description of the most Profitable Species of Swine* . . 12

CHAP. III. *Of the Different Species of Swine in Britain* 24

CHAP. IV. *The Difference of Gain by taking one Litter of Pigs from a Sow and then Feeding, from that of Feeding only, and pointing out the season such Pigs should be Farrowed* 30

CHAP. V. *The Management that ought to be observed as to the Sow, when with Young and giving Suck* 34

CHAP. VI. *The most proper and Cheapest Food for Swine, from being taken from their Dam until fit for Slaughter* 36

CHAP. VII. *The time the Breeding Sow ought to be put to the Boar, so as to bring, at proper seasons, two Litters within the year* 42

CONTENTS.

	Page
CHAP. IX. *Description of the best Conveniencies for Keeping Swine*	45
CHAP. X. *Proper Directions for Killing, Cutting up, and Curing and Drying the Flitches and Hams, for the London Market*	49
CHAP. XI. *A Description of Pickling and Kitting Pork, &c.*	54
CHAP. XII. *Of the Making of Brawn. Receipt for Preventing Bacon from turning Rusty. How to Cure the Measles, and the Rupture, in Pigs*	58
CHAP. XIII. *On the Supply of Fresh Pork throughout the Year to Private Families, at a small expence*	61
CHAP. XIV. *A Statement of the number of Swine fed in Annandale yearly, their average weight, and value. Importance of the Pork Trade to Bankers*	63
CHAP. XV. *On the Breeding of Swine in the Highlands and Western Isles of Scotland*	66
CHAP. XVI. *The advantages the Farmers along the Frith of Forth, and the neighbouring Counties, have over those of Dumfriesshire, for Breeding and Feeding Swine, and Carrying on the Trade of Curing Bacon*	71
CHAP. XVII. *Practical Remarks in addition to those already offered*	78
CHAP. XVIII. *Of the Prejudices so unjustly entertained against this Species of Stock*	80

PART II.

CHAP. I. *Recommendation to Gentlemen and Farmers to encourage the growth of Hemp in this Kingdom, with some Hints on its Culture and Management*	87

CONTENTS.

	Page.
CHAP. II. *On the best method of Harvesting Corn in a wet and backward season*	93
CHAP. III. *On the great saving which would accrue to the Farmer by employing in part Oxen, in labouring the Farm, and especially in working the Thrashing Machine*	96

A FEW HINTS AND RECEIPTS, FOR PREVENTING AND CURING DISEASES IN HORSES, CATTLE, AND SHEEP . 107

Of Sores and Bruises	108
On the use of Laxatives and Purgatives for Horses	116
On the Cure of Gripes in Horses	126
Receipt for Curing a Surfeit or Cough Bad Coat in Horses	133
Receipt for the Garget, or Udder Clap, to which Cows are subject after being put from giving Milk	ib.
For Curing the Redwater in Cattle	134
For the Foot Rot in Sheep	135
How to Prevent Sheep from Catching Cold after having been Shorn	137
For Curing and Preventing the Scab in Sheep	ib.
To Destroy Maggots in Sheep	139

INTRODUCTION.

All departments of husbandry require knowledge and care in conducting them. Some of them, however, are attended with more labour and expence than others, and also more slowly return to the husbandman a profit for his skill and capital employed in them. This observation particularly holds good in what relates to the management of our domestic animals; for of such some are much less expensive in their first cost and after maintenance, and yet become more rapidly productive, than others. No sort of live-stock will, in a given time, and with a small outlay, make a greater return than swine. Every man that has a house and yard may derive benefit by having some of them, providing the number kept be not disproportioned to the means possessed for comfortably subsisting them. —In this small Treatise, I shall endeavour to give some information regarding the breeding, rearing, and fattening, of this kind of stock, chiefly calculated for the use of common farmers, and of others who live in country situations, where potatoes can be raised for their food. If the observations here submitted about the proper method of keeping this sort of stock, shall prove of use to individuals or the public, or shall induce farmers in proper situations to engage in a lucrative branch of business, and shall gain to

INTRODUCTION.

the community a better supply than yet possessed of those valuable articles pork and bacon, or even if any of the hints given shall tend in the least degree to render more comfortable the condition of the poorer class of farmers and of cottagers, I shall be amply satisfied, and think the time spent in putting them together well employed.

The remarks on other subjects which are subjoined in the latter part of this publication, particularly on the treatment of wounds, and on one of the most common disorders of horses, &c. are meant to be useful to the same class of husbandmen, many of whom are unable, from the secluded situation in which they live, or from the narrowness of their circumstances, to obtain more perfect information on the subjects treated of, or to procure foreign assistance in the case of accident or disease.

Broomhill, near Annan,
13th June, 1811.

A TREATISE ON THE BREEDING OF SWINE, &c.

PART I.

CHAP. I.

Of the Natural History of the Hog.

BEFORE we proceed to the chief design of this work, which is to treat of the *Breeding of Swine*, the following account of the natural history of the animal may not be unacceptable to the reader.

To facilitate the study of Natural History, its various objects have been divided into classes, by selecting the properties which are common to many. As Nature herself, however, has drawn no very exact line of distinction between the different species of animals, it is often difficult to determine to which class some animals belong, or which they most resemble;

of this the hog, or swine, is a remarkable example. It is like the animals of the horse kind in the number of its teeth; in the length of its head; and in having but a single stomach: It is like the animals of the cow kind in its cloven hoofs, and in the position of the intestines; and it is like the animals of the claw footed kind, as the lion, the cat, and the dog, in its appetite for flesh, in not chewing the cud, and in its numerous progeny.

The animals of the hog kind, therefore, possess in the scale of existence a situation intermediate between those that live upon flesh, and those that live upon grass, being ravenous like the one and inoffensive like the other. Like the rapacious kinds, they are found to have short intestines; their hoofs also, though cloven to the sight, will, upon anatomical inspection, appear to be supplied with bones like beasts of prey; and the number of their teats also increase the similitude: on the other hand, in a natural state they live upon vegetables, and seldom seek after animal food, except when urged by necessity. They offend no other animal of the forest, at the same time that they are furnished with arms to terrify the bravest.

The wild boar, which is the original of all the varieties we find in this creature, is by no means so stupid nor so filthy an animal as that we have reduced to tameness; he is much smaller than the tame hog, and does not vary in his colour as those of the domestic kind do, but is always found of an iron grey, inclining to black; his snout is much longer than that of the tame hog, and the ears shorter, rounder, and black; of which colour are also the feet and the tail. He roots the ground in a different manner from the common hog; for as this turns up the earth in little spots here and there, so the wild boar plows it up like a furrow, and does irreparable damage in the cultivated lands of the farmer. The tusks also of this animal are larger than in the tame bred, some of them being seen almost a foot long. These, as is well known, grow from both the under and upper jaw, bend upwards circularly, and are exceeding sharp at the points. They differ from the tusks of the elephant in this, that they never fall; and it is remarkable of all the hog kind, that they never shed their teeth as other animals are seen to do. The tusks of the lower jaw are always most to be dreaded, and are found to give very terrible wounds.

The wild boar can properly be called neither a solitary nor a gregarious animal. The three first years the whole litter follows the sow, and the family lives in a herd together. They are then called beasts of company, and unite their common forces against the invasions of the wolf, or the more formidable beasts of prey. Upon this their principal safety while young depends, for when attacked they give each other mutual assistance, calling to each other with a very loud and fierce note; the strongest face the danger; they form a ring, and the weakest fall into the centre. In this position few ravenous beasts dare venture to attack them, but pursue the chace where there is less resistance and danger. However, when the wild boar is come to a state of maturity, and when conscious of his own superior strength, he then walks the forest alone, and fearless. At that time he dreads no single creature, nor does he turn out of his way even for man himself. He does not seek danger, and he does not much seem to avoid it. He does not shun the combat even with the lion, if provoked; he does not seek him to attack, but will not fly at his approach; he waits the onset of the lion, which he seldom makes unless compelled by hunger, and then exerts all his strength, and is sometimes successful. We are told of the combat of a lion and a wild boar, in a meadow near Algiers, which continued for a long time with incredible obstinacy. At last, both were seen to fall by the wounds they had given each other; and the ground all about them was covered with their blood.

This animal is therefore seldom attacked but at a disadvantage, either by numbers, or when found sleeping by moon-light. The hunting the wild boar is one of the principal amusements of the nobility in those countries where it is to be found. The dogs provided for this sport are of the slow heavy kind. Those used for hunting the stag, or the roe-buck, would be very improper, as they would too soon come up with their prey; and, instead of a chace, would only furnish out an engagement. A small mastiff is therefore chosen; nor are the hunters much mindful of the goodness of their nose, as the wild boar leaves so strong a scent that it is impossible for them to mistake its course. They never hunt any but the largest and the oldest, which are known by their tracks. When the boar is *rear'd*, as is the expression of driving him from his covert, he goes slowly and uniformly forward, not much afraid, nor very far before his pursuers. At the end of every half mile, or thereabouts, he turns round, stops till the hounds come up, and offers to attack them. These, on the other hand, knowing their danger, keep off, and bay him at a distance. After they have for a while gazed upon each other, with mutual animosity, the boar again slowly goes on his course, and the dogs renew their pursuit. In this manner the charge is sustained, and the chace continues till the boar is quite tired, and refuses to go any farther. The dogs then attempt to close in upon him from behind; those which are young, fierce, and unaccustomed to the chace, are generally the foremost, and often lose their lives by their ardour. Those which are older and better trained are content to wait until the hunters come up, who strike at him with their spears, and, after several blows, dispatch or disable him. The instant the animal is killed, they cut off the testicles, which would otherwise give a taint to the flesh; and the huntsmen celebrate the victory with their horns.

The hog, in a natural state, feeds chiefly upon roots and vegetables; it seldom attacks any other animal, being content with such provisions as it can procure without danger. Whatever animal happens to die in the forest, or is so wounded that it can make no resistance, becomes a prey to the hog, who seldom refuses animal food, how putrid soever, although it is never at the pains of taking or procuring it alive. For this reason, it seems a glutton rather by accident than choice, content with vegetable food, and only devouring flesh when pressed by necessity, and when it happens to offer; and though, in its domestic state, it seems the most sordid and brutal animal in nature, devouring indiscriminately every thing that comes in its way, yet, in its wild state, it is of all other quadrupedes the most delicate in the choice of what vegetables it shall feed on, and rejects a greater number than any of the rest. The cow, for instance, as we are assured by Linnæus, eats two hundred and seventy-six plants, and rejects two hundred and eighteen; the goat eats four hundred and forty-nine, and rejects an hundred and twenty-six; the sheep eats three hundred and eighty-seven, and rejects an hundred and forty-one; the horse eats two hundred and sixty-two, and rejects two hundred and twelve; but the hog, more nice in its provision than any of the former, eats but seventy-two plants, and rejects an hundred and seventy-one. In the orchards of peach-trees in North America, where the hog has plenty of delicious food, it is observed, that it will reject the fruit that has lain but a few hours on the ground, and continue on the watch whole hours together for a fresh windfall.

However, the hog is naturally formed in a more imperfect manner than the other animals that we have rendered domestic around us, less active in its motions, less furnished with instinct in knowing what to pursue or avoid. The coarseness of its hair, and the thickness of its hide, together with the thick coat of fat that lies immediately under the skin, render it in some degree insensible to blows, or rough usage. Its other senses seem to be in tolerable perfection; it scents the hounds at a distance; and, as we have seen, is not insensible in the choice of its provisions. When the wind blows with any vehemence, it is so agitated as to run violently towards its sty, screaming horribly at the same time, which seems to argue that it is naturally fond of a warm climate. It appears also to foresee the approach of bad weather, bringing straw to its sty in its mouth, preparing a bed, and hiding itself from the impending storm. Nor is it less agitated when it hears of any of its kind in distress: when a hog is caught in a gate, as is often the case, or when it suffers ringing or spaying, all the rest are then seen to gather round it, to lend their fruitless assistance, and to sympathize with its sufferings. They have often also been known to gather round a dog that had teazed them, and kill him upon the spot.

Most of the diseases of this animal arise from intemperance; measles, imposthumes, and scrophulous swellings, are reckoned among the number. It is thought by some that they wallow in the mire to destroy a sort of louse or insect that is often known to infest them. They are generally known to live, when so permitted, to eighteen or twenty years; and the females produce till the age of fifteen. As they produce from ten to twenty young at a litter, and that twice a year, we may easily compute how numerous they would shortly become, if not diminished by human industry. In the wild state they are less prolific; and the sow of the woods brings forth but once a year, probably because exhausted by rearing up her former numerous progeny*.

The wild boar was formerly a native of Great Britain, as appears from the laws of Hoeldda, the famous Welsh legislator, who permitted his grand huntsman to chace that animal from the middle of November to the beginning of December. The vast forest that grew on the north side of London, was the retreat of many fallow deer, wild boars, and bulls. The Caledonian forest, supposed to be Ettrick forest, was anciently the retreat of the Caledonian boars, which were remarkable for their fierceness, and which are now to be no where met with in Britain. Inglewood forest, betwixt Carlisle and Penrith, was once the retreat of the wild boar. During the day, he commonly remained in the most sequestered part of the wood, and came out in the night in quest of food.

The snout of an old boar is said to be the only part that is esteemed†, but every part of the castrated

---

* Goldsmith's History of the Earth and Animated Nature.

† At a time when fresh meats were seldom eaten, brawn was considered as a great delicacy; the boar's head soused was anciently the first dish on Christmas-day, and was carried up to the principal table in the hall with great state and so-

and young boar, not exceeding a year old, makes delicate eating. The ancients castrated young boars,

---

lemnity. Hollingshed says, that in the year 1170, Henry I., upon the day of the young prince's coronation, served his son at table as sewer, bringing up the boar's head with trumpets before it, according to the manner. There is also a singular ceremony relating to the boar's head, still retained at Queen's College, in Oxford. For this indispensible ceremony, as also for others of that season, there was a Carol, which Wynkyn de Worde has given us as it was sung in his time, with the title, ' *A Carol bringing in the Boar's Head.*'

> *Caput Apri defero*
> *Reddens Laudes Domino.*
> The boar's head in hand bring I,
> With garlandes gay and rosemarye;
> I pray you all sing merrely,
> *Qui estis in convivio.*
>
> The boar's head, I understande,
> Is the chief servyce in this lande,
> Looke wherever it be fonde,
> *Servite cum cantico.*
>
> Be gladde lordes, both more and less,
> For this hath ordained our Stewarde,
> To chere you all this Christmasse,
> The boar's head with mustarde.

This Carol, says Mr Warton, is still retained at Queen's College. There is indeed in the college an old legend, that a wild boar which infested the neighbourhood of Oxford, was killed by a taberdar of this college on Christmas-day, as he was going to serve a church; and that he killed it by thrusting his copy of Aristotle down the throat of the animal, protecting his arm with some part of his gown. This story, it is pro-

which they carried off in the absence of the old ones, and returned them back to the woods, where they

---

bable, may have contributed to the continuance of the ceremony of the boar's head at Queen's College longer than any where else. The song, however, has no allusion to it; it simply states, that the boar's head is the rarest dish in all this land, and that it has been provided in honour of the King of Bliss.

There is a song on this supposed feat of the taberdar, written by the present Dr Harrington of Bath, and printed in the ' Oxford␣ausage,' so full of wit and humour, that we assure ourselves our readers will not be displeased to find it annexed to this note.

IN HONOUR OF THE CELEBRATION OF THE BOAR'S HEAD,

AT QUEEN'S COLLEGE, OXFORD.

I sing not of Roman or Grecian mad games,
The Pythian Olympic, and such like hard names;
Your patience a while, with submission, I beg,
Whilst I study to honour the feat of Cool Reg.
   Derry down, down, down, derry down.

No Thracian bowls at our rites e'er prevail,
We temper our mirth with plain sober mild ale;
The tricks of old Circe deter us from wine,
Though we honour a boar we won't make ourselves swine.
   Derry down, &c.

Great Milo was famous for slaying his ox,
Yet he proved but an ass in the cleaving of blocks;
But we had an hero for all things was fit,
Our motto displays both his valour and wit.
   Derry down, &c.

grew fat, and their pork was much better than that of the common hogs.

    Stout Hercules labour'd, and look'd mighty big,
    When he slew the half-starved Erymanthian pig:
    But we can relate such a stratagem taken
    That the stoutest of boars could not save his own bacon.
        Derry down, &c.

    So dreadful this bristle-back'd foe did appear,
    You'd have sworn he had got the wrong pig by the ear,
    But instead of avoiding the mouth of the beast,
    He rammed in a volume, and cried *Græcum est.*
        Derry down, &c.

    In this gallant action such fortitude shewn is,
    As proves him no coward, nor tender Adonis;
    No armour but logic, by which we may find
    That logic's the bulwark of body and mind.
        Derry down, &c.

    Ye squires, that fear neither hills nor rough rocks,
    And think you're full wise when you outwit a poor fox,
    Enrich your poor brains, and expose them no more,
    Learn Greek, and seek glory from hunting the boar.
        Derry down, &c.

CHAP. II.

The Rise and Progress of Manufacturing Pork into Flitches and Hams in Dumfriesshire, and sending it to the London Market, with a Description of the most Profitable Species of Swine.

WITHOUT attempting to state the precise time when swine first became domestic animals in this kingdom, I shall only show when they began to multiply in Dumfriesshire so as to become an article of commerce.

In many districts of Britain, swine were kept in large herds at a very early period. So far back as 2673 years ago, it appears that a numerous herd of them was kept by a British prince in the neighbourhood of Bath; to whom, if we can believe the following story, we owe the discovery of the medicinal qualities of the hot springs there.

Baldred, eldest son of Lud Hudibras (then king of Britain), it is said, having spent eleven years at Athens in study came home leprous, and was in consequence confined, to prevent the infection of his disease. Having effected his escape, however, he went very remote from his father's court, into an untravelled part of the country, and offered his service in any common employment. He entered into service at Learwick, a small village three miles from Bath, where his business was to take care of pigs, which he was to drive from place to place for their advantage in feeding upon acorns, haws, &c. While at his usual employment one morning, part of the drove of swine, as if seized with a frenzy, ran down the side of the hill into an elder moor, till they reached the spot of ground where the hot springs of Bath now boil up, and from thence returned covered with black mud. The prince being of a thoughtful turn, and very solicitous to find out the reason why the pigs, that wallowed in the mire in summer to cool themselves should do the same in winter, at length perceived a steam arise from the place where they had wallowed, and making his way to it, found it warm. Having thus satisfied himself that it was for the benefit of the heat that the pigs resorted thither, he observed that after a while they became whole and smooth from their scurfs and eruptions by often wallowing in this mud. Upon this he considered within himself why he should not receive the same benefit by the like means; he tried it with success, and finding himself cured of his leprosy declared who he was. His master, though incredulous at first, being at last persuaded to believe him, went with him to court, where he was owned, and upon succeeding his father he erected the baths*.

It appears, that in the Highlands of Scotland large

---

* In one of these baths there is still seen a statue of King Baldred, which was erected in 1699, under which is the following inscription on copper. 'Baldred, son of Lud Hudibras, eighth king of the Britons from Brute, a great philosopher and mathematician, bred at Athens, and recorded the first discoverer and founder of these baths, 863 years before Christ.'

flocks of swine have been kept and reared up in the glens, in the same manner as their horses, cattle, and sheep. About forty-five years ago, large droves of Highland swine were brought into Annandale and sold to the farmers, who bought them more from motives of curiosity than of profit. They were very small, with long bristles upon their backs; and were sold, when one year old or upward, for four or five shillings per head. I can very well remember of people running from all quarters to see them. I understand the flocks of swine are not now so numerous in the Highlands as formerly. But more of this afterwards.

Great herds of swine are kept at present in the county of Northumberland, and I have frequently seen farmers, at Ovingham and Newcastle fairs, selling shots (as they are so called when they are six or eight months old), from twenty to thirty in a drove. Jobbers from Dumfriesshire commonly purchased great numbers of them after harvest and brought them thither, where they found a ready market among the farmers, who fed them that season. Notwithstanding the immense number of breeding sows that are kept in the county of Dumfries, many hundreds of pigs and shots are brought from England and fed in Scotland. In all the corn districts of Yorkshire, an immense quantity of swine are kept; it is very common to see a person attending thirty, forty, or fifty swine, upon clover or stubble fields; and in Norfolk and the adjoining counties herds of the same kind are kept. I once travelled some miles with a farmer to view his stock of swine, which amounted to three hundred. The late and wet harvest having spoiled his barley crop, this induced him to feed more than his ordinary number of them. His conveniencies for them, therefore, were not so complete as one would have wished. In Norfolk the swine are not of so large a breed as in this country, they weigh only from six to ten stone; they are mostly used in pork. The intelligent farmers there say, that if it were not for their swine, they could neither feed their families nor pay their rents.

But though swine were kept so early in the south of England, it appears from the following anecdote that they were very little known in the north of England and south of Scotland.

Within the last century (probably about ninety years ago) a person in the parish of Ruthwell, in Dumfriesshire, called the 'Gudeman o' the Brow,' received a young swine as a present from some distant part; which, from all the information I could get, seems to have been the first ever seen in that part of the country. This pig having strayed across the Lochar into the adjoining parish of Carlavroc, a woman who was herding cattle on the marsh, by the sea side, was very much alarmed at the sight of a living creature, that she had never seen nor heard of before, approaching her straight from the shore as if it had come out of the sea, and ran home to the village of Blackshaw screaming. As she ran, it ran snorking and grunting after her, seeming glad it had met with a companion. She arrived at the village so exhausted and terrified, that before she could get her story told she fainted away. By the time she came to herself a crowd of people had collected to see what was the matter, when she told them, that 'There was a diel came out of the sea with two horns in his head (most likely the swine had pricked ears) and chased her, roaring and gaping all the way at her heels, and she was sure it was not far off.' A man called Wills Tom, an old schoolmaster, said if he could see it he would 'cunger the diel,' and got a bible and an old sword. It immediately started up at his back and gave a loud grumph, which put him into such a fright that his hair stood upright in his head, and he was obliged to be carried from the field half dead.

The whole crowd ran some one way and some another; some reached the house tops, and others shut themselves in barns and byres. At last one on the house top called out it was 'the Gudeman o' the Brow's grumphy,' he having seen it before. The affray was settled, and the people mostly reconciled, although some still entertained frightful thoughts about it, and durst not go over the door to a neighbour's house after dark without one to set or cry them. One of the crowd who had some compassion on the creature, called out, 'give it a tork of straw to eat, it will be hungry.'

Next day it was conveyed over the Lochar, and it seemed to find its way home. It being near the dusk of evening, it came grunting up to two men pulling thistles on the farm of Cockpool. They were much alarmed at the sight, and mounted two old horses they had tethered beside them, intending to make their way home. In the mean time the pig got between them and the houses, which caused them to scamper out of the way and land in Lochar moss, where one of their horses was drowned, and the other with difficulty relieved. The night being dark, they durst not part one from the other to call for assistance, lest the monster should find them out and attack them singly; nor durst they speak above their breath for fear of being devoured. At day-break next morning they took a different course, came by Cumlongon castle and made their way home, where they found their families much alarmed on account of their absence. They said that they had seen a creature about the size of a dog, with two horns in its head, and cloven feet, roaring out like a lion, and if they had not galloped away, it would have torn them to pieces. One of their wives said, ' Hout man, it has been the Gudeman of the Brow's grumphy; it frightened them a' at the Blackshaw yesterday, and poor Meggie Anderson maist lost her wits, and is ay out o' ae fit into anither sin-syne.'

The pig happened to lie all night among the corn where the men were pulling thistles, and about day-break set forward on its journey for the Brow. One Gabriel Gunion, mounted on a long tailed grey colt, with a load of white fish in a pair of creels swung over the beast, encountered the pig, which went nigh among the horses feet and gave a snork. The colt, being as much frightened as Gabriel, wheeled about and scampered off sneering, with his tail on his riggin, at full gallop. Gabriel cut the slings and dropt the creels, the colt soon dismounted his rider, and going like the wind, with his tail up, never stopped till he came to Barnkirk point, where he took the Solway Frith and landed at Bowness, on the Cumberland side. As to Gabriel, by the time he got himself gathered up, the pig was within sight, he took to his heels, as the colt was quite gone, and reached Cumlongon wood in time to hide himself, where he staid all that day and night, and next morning got home almost exhausted. He told a dreadful story! The fright caused him to imagine the pig as big as a calf, having long horns, eyes like trenchers, and a back like a hedge-hog. He lost his fish, the colt was got back, but never did more good, and as to Gabriel, he soon after fell into a consumption, and departed this life about a year after.

About this time also a vessel came to Glencaple quay, a little below Dumfries, that had some swine on board, most likely for the ship's use; one of them having got out of the vessel in the night, was seen on the farm of Newmains next morning. The alarm was spread, and a number of people collected. The animal got many different names, and at last it was concluded to be a brock. Some got pitchforks, some clubs, and others old swords, and a hot pursuit ensued; the chace lasted a considerable time, owing to the pursuers losing heart when near their prey and retreating. Robs Geordy, having rather a little more courage than the rest, ran 'neck or nothing,' forcibly upon the animal, and run it through with a pitchfork, for which he got the name of 'stout hearted Geordy' all his life after. There is an old man, nearly a hundred years of age, still alive in the neighbourhood where this happened, who declares that he remembers of the Gudeman of the Brow's pig, and the circumstances mentioned; and he says it was the first swine ever seen in that country.

About the year 1760, one would scarce have seen twenty swine in a parish throughout Dumfriesshire; but about 1770, they began to be more plentiful, and every farmer kept one or two. A market was opened at Longtown, in Cumberland, about that time, or a little earlier, and a few dead carcases exposed for sale every Thursday in each week during the season. Dumfriesshire may long remember the late Lord Graham of Netherby for the institution and support of that market, it was the sole cause of the progress of the trade, and its arriving at such perfection, as there was no other market for the swine fed in that quarter: farmers at the distance of thirty miles sent their swine to it, and had always the ready money home with them. The market could not have been begun or carried on but for the liberality of Lord Graham, who allowed his tenants to retain their Lammas and Candlemas rents in their hands for the purchase of pork, which they manufactured into bacon and hams for the London market. This occasioned a circulation of money in the neighbourhood, which at that period was very scarce and much wanted. Such goodness and regard for the interest of his tenants and the public welfare, shewed a liberal mind, and the virtuous example of his Lordship does not seem to be lost, as his son and heir, Sir James Graham, Bart. of Netherby, has lately shewed a similar generosity for his tenants, which is equally praiseworthy. There is still a considerable market at Longtown, and several farmer in the low parts of Annandale send their pork to it.

I lately spoke with an old farmer, who told me, that forty years ago, he took four swine to Longtown and sold them at 2s. 6d. per stone*. At that period there were hardly any curers of bacon in Annandale; soon after that some people began to attempt curing hams. I remember one Betty Little, in the parish of Middlebie, trying to cure a pair of hams of her father's swine; she succeeded, and was afterwards sent for to all parts of the country to cure hams. A number of the old people thought it was witchcraft; for they could not understand how she could cure them with the bone in. Indeed, of what the farmers cured for their own use, they always took out the ribs, and cut out the shoulder bone. Simeon Johnston, in the parish of Middlebie, was the first that commenced the trade of purchasing and curing hams in Annandale. The farmers of the parish of Hoddam took their hams to

---

* A stone 14 lb., 16 oz. to the lb.

him, some at the distance of ten miles. The price then was 2½d. or 3d. per pound. A little after this some others began to try the curing of both flitches and hams, and as the demand increased, the farmers augmented their stock, and betwixt the years 1775 and 1780 it became a considerable trade; and every person who could raise a little money laid it out in the winter season upon pork. In the year 1790 it became a general business, the purchasers travelling through the country and buying the swine alive, and some by weight, to be sent to their houses.

About this time a market was established at Ecclefechan; at which there was sold in the first season nearly £.7000 worth. The price then was from 3s. 6d. to 4s. per stone; but that market dwindled away, owing to the author, the founder of the market, leaving the town. Since that time a market has been established at Lockerby, where a great number of dealers still attend. About ten years ago the number of pork carcases offered for sale at this market weekly, from Martinmas till the 1st of April, was from 50 to 110, average 80, weighing from 12 to 24 stones each. The number is greatly increased of late years, I have not the smallest doubt that upwards of £.9000 worth are sold there every season*. Another still more considerable market for pork is now held at Dumfries, where the Annandale curers meet the Gal-

---

* I feel myself indebted to William Stewart, Esq. of Hillside, for the pains he has taken in giving me the information about that market.

loway farmers. A weekly market for pork was established at Annan about three years ago. The magistrates of that place, very generously, give upon every market day a premium of five shillings for the best fed pig, and a like sum to the person who exposes for sale the greatest number of swine. These premiums are given by judges chosen for that purpose, and paid regularly every week during the season, which lasts about five months. This liberal encouragement seems already to have had an effect, and probably will be the means of soon rendering this market the largest in the county. The magistrates of Lochmaben have adopted a similar method to those of Annan for the establishment of a pork market, which, I understand, is likely to succeed. There is now a branch of the Commercial Bank of Scotland at Annan, which will greatly increase that market, and be of infinite benefit to the bankers, as the circulation of notes in that trade is preferable to any other; the agent being besides respectable, and well acquainted with country business.

At what period the curing of bacon commenced in Yorkshire and Westmoreland I am quite unable to determine, but it would appear to have been at a much earlier date than in the north of England and south of Scotland, for the people in London are so ignorant to this day as to imagine all the hams are from Yorkshire and Westmoreland. The county of Dumfries now sends six times the quantity that the former does. It is also become a general trade all over Cumberland, as well as Dumfriesshire and Galloway. The spark of this trade was first kindled in Yorkshire, and seems now to be spreading northward. I hope this little treatise will prove of use by conveying information to the northern part of Scotland.

In this country the different species of swine are so blended and intermixed one with another, that it is nearly impossible to mention any particular breed. The Berkshire pig is generally allowed to be a good kind, and is, I believe, preferable to many. They are generally of a brown, or rather reddish colour, with black spots; the ears bending forward, but not hanging down so much as those of the large Cumberland kind; short legged; small boned, and very easily fattened. Mr Culley mentions one of this breed which was killed in Cheshire, that measured nine feet eight inches from the nose to the end of the tail, and four feet and a half in height, and weighed when killed eighty-six stone eleven pounds avoirdupois.

There is none of the real Berkshire breed in Scotland at present, and we must therefore resort to that species which can be most easily got, until some spirited agriculturist introduce them into the north of England and Scotland, which I should be happy to see. I can, however, recommend a cross from the best shaped and largest species of the black kind commonly called the Guinea breed, with a white Cumberland boar. A second or third cross from this boar will answer best, and will weigh from eighteen to twenty-two stone, when from fourteen to eighteen months old. Three swine of this description will feed upon the same quantity of food as two of the large long legged kind will require. The latter cannot be brought to market so young as the former; for the coarser the breed of any animal, the more time it will require to be ready for the market. A bad sort causes the feeder to lie longer out of his money, costs more in feeding, and is not so marketable when fat. The thick easily fed swine, are from 1s. to 2s. per stone more valuable than the large thin ones.

If the above described stock, should not please, nor answer the expectation of the breeder, he will not be much disappointed, as, in the course of the year, he can feed off his whole stock, and procure another more to his liking. A change of the breed of swine is more easily effected than that of either horses, cattle, or sheep.

CHAP. III.

Of the Different Species of Swine in Britain.

The Chinese, or black breed, are now common in Britain. They have short legs, are smaller, and their flesh whiter than the common kind. It is said this species are found in Guinea, and that they are very numerous in the Friendly, Society, and many of the other newly discovered islands in the South Sea.

The real Highland hog is well adapted to the climate where it is found; it is well shaped, the bone small, and the ham plump; and when well fed is delicious eating. A cross between them and the Chinese breed, I apprehend, would answer well, as this would increase the weight of the Highland hog and add to the hardiness of the Chinese, and be the means of lessening the bone of the latter; a heavy carcase upon a small bone is the most proper animal for feeding, and this sort would be fattened with little food.

There is a species of swine in Ross-shire, mostly black spotted, and of a very plump make, which seem to be easily fattened, and get early to a proper size for porking. It is much superior to that diminutive creature that is so common in Fifeshire and the Lothians, which is neither easily fattened nor ever arrives to any considerable weight. It seems to be a town talk when a hog pig of the latter species weighs eight stone Dutch; perhaps an old sow may chance to arrive at 14 or 16 stone. The breeders in these counties must eradicate such a mongrel kind before they can do any good by keeping swine.

I have lately seen hogs of a black colour, at the Earl of Moray's, at Dennibirsel, that have close feet like a horse, instead of being cloven footed; but they seem to have nothing else particular about them to recommend them beyond other swine in that neighbourhood. The conveniencies in which they are kept are extensive, but of a bad construction, not being of modern contrivance. These erections might be very much improved at a small expence, as there are materials upon the spot.

From Dunbar to Berwick the swine seem to be of a better sort, and will weigh more, but that may be owing to their superiority in feeding, as the practice of kitting pork for the London market is carried on at Berwick with great spirit.

In Northumberland they have a kind of black spotted swine, not very large, which makes good bacon and pork: they are not a pure breed, but rather of a mongrel kind.

The Cumberland breed are generally large, and mostly of a white colour, many of them would feed from 25 to 30 stone, when 14 or 18 months old. They are strong boned, with large ears hanging down over their eyes so far as their snout. In that country families live much upon bacon, and it is very common to see three or four of these heavy swine hanging in a kitchen for family use. Almost every farmer keeps one or two breeding sows; they pay great attention to feeding them properly, and sell their sucking pigs at market towns when from six to eight weeks old; they are generally very large and fat, and will sell at from 12s. to 20s. each. It is common with them to give their pigs new milk from the cow twice a-day the last three weeks they are upon the sow, as they then get strong and their dam cannot support them, at least to make them so fat as they are generally shown in the market.

There will be about 120 pigs sold in Carlisle market

every Saturday through the season. Most of them are bought by Dumfries Jobbers, and taken to Dumfries market in carts on the Wednesday, where they are sold to the Galloway farmers for feeding. So foolish are the cottagers on the borders of Scotland, that they always purchase the largest pig they can get. I am certain it would be more for their interest to purchase pigs of a smaller breed, as the large swine require much more food.

In Westmoreland their swine are not of so large a breed, this district not being so much of a corn country; they are, however, carefully fed, and reared to a good weight.

In Lancashire they keep a good sizeable swine, which is mostly black and white spotted; they feed to about 20 stone. In this county they feed mostly with beans, and other substantial food, and their pork is very firm. I have known them feed to such a degree that the animal could scarce eat any thing; they would bake the oatmeal into little balls, and when it began to loathe that feeding give it new milk from the cow, often making it so fat that it could not rise. They manufacture very little into bacon, being mostly all sold in the market, by the feeder, in sides and quarters, for people to cure and dry for family use.

In Cheshire they keep a very proper sort of pigs, they are generally black spotted, with pretty large ears, which bend forward but do not hang so straight down as in the Cumberland hog; with short legs and broad back. Here they feed to great perfection, and pay the utmost attention to the animal; they are as regular in the cleaning and feeding of their hogs as with their horses. Cheshire being a great cheese country they have plenty of whey and other offals from the farm, which they keep carefully for their pigs. These hogs seem less adulterated than those in Scotland, every one being very much like another. They generally weigh from 20 to 25 stone.

I would recommend it to some of the principal cheese-makers in the west of Scotland, particularly in Ayrshire, to make a tour through part of England, and pay attention to the management of swine in Cheshire. They would then know the value of whey, and how to use it; their pains and expence would not be lost, as the cheesing districts in Ayrshire are as well calculated for keeping pigs as Cheshire.

Great numbers of pigs are fed in North Wales. I once happened to be at Mould fair, in Flintshire, where there might be 500 fat swine shown for sale. Jobbers and butchers from Manchester and its neighbourhood purchased large droves of them, and they were all sold off; the average weight might be 14 stone. They were all of a white colour. I was very much pleased to see so many plump, well feed animals, all so much alike that one would have imagined they were all from one sow. Most of them were fed in Denbighshire and Carnarvonshire. In South Wales also, many swine are kept, but I never had an opportunity of viewing them. But it is needless to mention a county here and there, for over all England and

Wales swine are kept in large flocks, and are found of the greatest benefit to the community at large.

Of all the sorts of hogs I have seen, however, and that is not a few, I would give the preference to the Cheshire, or rather the Berkshire hog; this kind will grow to a good size, is easily fed, small boned, and of a handsome make. Many of them are of a singular reddish colour, and I have seen droves of them that when the sun shone on them appeared of the colour of gold. I wish very much to see them introduced into Scotland, and am certain they would be more profitable than any other species.

Many of our gentlemen put their money to a worse use than if they were to send a careful person, that has a knowledge of swine, to bring a small drove of sows, and a few boars, of the true Berkshire breed. They can be got in Cheshire, or the neighbouring county, without going so far as Berkshire. Indeed it would not be spending their money without reward, for they would be amply repaid their trouble, besides doing an infinite benefit to the community. If ever the pork trade take place in Fifeshire, or in the Lothians, the breed ought to be changed, as the present stock is entirely out of the question. The journey may be thought long for the purpose of bringing swine, but time and perseverance will accomplish it, and the example must be set by some individuals introducing a proper stock, before the prejudices naturally operating against any new plan can be removed. It would be a good scheme for a number of farmers to join in sending a man to purchase a sufficient number of sows and boars. If any thing of the kind take place, on application being made to the author at Leven, in Fifeshire, he will furnish them with a person fully able to execute the business, besides becoming one of the subscribers himself. The month of August would be the fittest time for this species of stock travelling.

CHAP. IV.

The Difference of Gain by taking one Litter of Pigs from a Sow and then Feeding, from that of Feeding only, and pointing out the season such Pigs should be Farrowed.

A FARMER that has 120 acres in crop, may, with advantage, feed twenty swine. The stock may be proportioned according to the size of the farm. Let the intended breeder and feeder purchase twenty sow pigs and one boar pig, which were farrowed the beginning of June. The June following they will all have pigs, which they will suckle about two months. These pigs must be all sold just when weaned, except twenty-one to keep up the stock, which must be all chosen of the handsomest shape, and good size, in order that the stock may continue uniform. The farmer will now be in a proper state to go on, and will have no

ON THE BREEDING OF SWINE.

occasion to lay out money for stock. After the pigs in question are sold, the sows must not go above one month at large before they are put up to feed; say about the beginning or middle of September. The boar must be kept from them until put up, so as they may be nearly half way gone with pigs when ready for killing, which they should be, as in this state they consume but little food comparatively speaking, are much inclined to sleep, and very quiet. If this plan is followed, the sow is nearly a third part of her time with young, and, consequently, a considerable saving to the feeder. If they are kept in good condition when giving suck, they will be fat in two months. It must always be a rule not to allow them the boar when poor, as they cannot be fat when half way gone with pigs. It will not answer to turn them out to the fields any length of time after the pigs are weaned, as they will all be a breaming, and troublesome to manage; otherwise I would have recommended the pigs to be farrowed a month or six weeks sooner, as they would then bring more money. In that case the sow would be so early put up to feed that she would be fattened before the season for curing bacon commenced. It may be observed, that a sow should always either be with young or giving suck, for if they are allowed to run a breaming, they will lose flesh in place of gaining it.

I shall now return to the pigs which were for sale. The 20 sows would bring at an average $7\frac{1}{2}$ pigs each, in all 150; after deducting 20 for keeping, 130 will remain for sale; and, as it is a proper season for selling them, they will bring 14s. per head, which amounts to £.91. This is a moderate estimate, considering the enormous price pigs have sold at lately. Now, as the sows will, when fed, be about eighteen months old, and will weigh, say 20 stone each, 400 at 6s. per stone, is £.120, which is now-a-days below an average price, as pork found a ready market last year at 8s. 6d. to 9s. per stone. The nearest calculation that I can make for the expence of keeping (not to enter into particulars) is £.4 per head; which, for 20 sows, is in all £.80: but I have always found, that the pigs for sale, with the boar when fat, will defray all expences, including the sow-keeper's wages, &c. not accounting any thing for the valuable dunghill: This makes a clear profit to the farmer of £.120, by taking one litter of pigs from a sow and then feeding. It may be mentioned, that a young boar of the same litter of the sow pigs is sufficient to propagate the stock.

We shall next view the difference of gain between the method of feeding only, and that of taking a litter of pigs before feeding, as above described. The former will not get to any greater weight at the same age, or bring more money when fat, than the latter. The reader may easily strike a balance betwixt the two methods. Allow £.4 per head to be deducted for the keeping of the feeding swine, and there will remain £.40 for the feeder; balance in favour of the breeding swine £.80. But even allowing the 20 feeding swine to make £.60 clear, after deducting the expence of keeping, the breeding sows have still a balance of £.60 in their favour, which is the difference of gain between taking a litter of pigs from a sow and then feeding, and that of feeding only.

In the foregoing calculation it might be supposed that the food for the swine is purchased, but this is not understood to be the case in rearing and feeding swine, as there is very little outlay of money in making them ready for market.

I do not at all wish that any person should adopt my plan on a large scale without giving the different methods a trial; one scheme will answer in one country, when it will not in another. Perhaps taking a litter of pigs from a sow will not answer in a district where so many pigs cannot be sold. In this case, keep ten pigs all of one litter; cut five and allow the remaining five to go uncut, from which take a litter of pigs. Allow the whole to be fed alike and kept to the same age, when the difference of value will be exactly known. The breeder and feeder must be regulated as to the number of sows he will take a litter from and then feed, by the situation of the place. If he cannot dispose of a large number of pigs, put fewer to breeding, and feed only. If this plan of breeding a litter and then feeding were to become general, there would be little or no sale for the pigs. The sale of pigs fluctuates according to the number bred: thus a farmer should look forward for his own government.

CHAP. V.

The Management that ought to be observed as to the Sow, when with Young and giving Suck.

A sow, when with young, ought not to be entirely confined to a hut, nor allowed to travel too much; but only to walk about at pleasure in the swine-yard, or any convenient place, where she can do no damage. When half way gone with pigs, she should not be allowed to sleep with a number of swine, at least not to lie throng, as she is but too apt to have her young killed with the feet of her bedfellows.

It may not be unnecessary to remark, that a sow goes with young 112 days, or sixteen weeks; not above a day over or under that time. A few hours previous to her farrowing, she will be observed carrying straw, &c. in her mouth, to make a bed with. I must recommend it to be carefully observed, not to have much litter or long straw in her apartment when bringing forth her young, and even for a few days after; as having too much roughness about her would endanger covering some of the pigs, so that their dam could not perceive where they were, and consequently lie down upon them, although she is very careful to push what part of them she sees before her, with her snout, before lying down.

She ought to be well fed when giving suck, as it is the only time she has a craving appetite for eating and drinking. This would keep both herself and family in good condition. If she is once allowed to get poor, it will take double the quantity of food and time to bring her to her former state.

If a sow happens to have a great number of pigs, it must be noticed, that in order to prevent her being brought to poverty, and having starved pigs, they must be fed two or three times a-day with milk and wheat bran. If the milk is scarce, mix a little water with it slightly warm, and take care that the mixture is not too thick. When the pigs are feeding thus, turn out the sow from them for an hour or two, which will enable her to gather some little milk; but she must never be allowed to be absent from her pigs all night. This custom ought to be adopted when a sow brings forth more young than she can support, which is sometimes the case. It has another good effect, when the pigs are accustomed to take meat, it will prevent their taking what is called the *weaning brash*.

The sow and pigs should lie dry and warm, and have plenty of *drinks* given them, mixed with a little meal, or wheat bran, warmed moderately. If the sow and pigs are well fed, the pigs will be ready for sale when six weeks old. If they are not well fed, they must suck eight weeks, which will be found a disadvantage to the breeder, and will cause the loss of two weeks extra food to both the dam and her family. Every breeder and feeder will find his own advantage in never allowing any of his tribe to go to bed with a hungry belly.

CHAP. VI.

The most proper and Cheapest Food for Swine, from being taken from their dam until fit for slaughter.

When the pigs are taken from the sow and weaned, whether with the breeder or purchaser, they must be regularly fed three times a-day, with a little wheat bran, barley dust, or meal of some description or other, mixed with water warmed a little, though not warmer than their mother's milk. In a few weeks after they are taken from their dams and fed thus, they will not be found so nice, but will occasionally amuse themselves with a few raw potatoes, turnips, or any kind of grain. Whoever has been accustomed to feed swine will not throw away the washings of a kitchen, whey, or butter milk, &c. which the kitchen maid ought to be made throw into a vessel together with any offals from the family, for the use of the pigs. The young stock, after the harvest is concluded, may be allowed to range at large among the fields to pick up any left corn, beans, or peas, &c. and their mothers may be allowed to go along with them, in case they are not put up to feed. The keeper must always be along with them, to prevent their doing any damage, or straying too far abroad. They may likewise be allowed to go upon the new sown wheat, provided the ground is dry, as their trampling is of some service to that grain. But it must then be observed, to have nose jewels well secured in their snouts, to deter them from turning up the ground. Before going out in the morning, always give them something for breakfast, and on their return in the evening have a similar allowance in their manger. A few Swedish turnips, or even common ones (if the other are not plenty,) cut into small pieces, may be given them; or potatoes occasionally, as the feeder may see requisite.

The pigs, at the first offer of turnips, may not be so fond of them as of potatoes. Mix them, therefore, in a trough with a few oats or barley, which withdraw by degrees, and they will soon be brought to make a meal of them. The prevailing practice of feeding swine, is to boil potatoes, and after draining the water from them, to mix them with oat or barley meal, which is given them in such proportions as the keeper may deem sufficient. My opinion is, that the potatoes would answer much better to be made ready by steam. Those who have not kettles, or caldrons, made on purpose, may have a fixed pot. In which case, take a barrel, or hogshead, and knock in the one end of it; bore a number of half inch holes in the bottom, so as to receive the steam from the pot. Let the vessel be proportioned to the size of the boiler, so as to receive the mouth of the pot within its edges. A piece of wrought clay may be used to join the pot and cask, to keep in the steam. A lid, or covering, will be necessary. The pot must be nearly full of water, and as many potatoes in the cask as occasion may require. Apply a fire to the pot sufficient to keep it boiling. The potatoes will as soon be ready for mashing down as if boiled in water; they will be quite dry, and keep longer without souring. The cask may be set upon the boiler when empty, and then filled again with potatoes; but it would be rather troublesome to take the potatoes out with the cask standing at such a height. To remedy this inconvenience, something should be contrived to lower the cask of boiled potatoes to the ground. A tackle, or pully, such as is used at warehouses or ships, may answer the purpose, if properly fixed above the pot; or a small door may be cut out of one side of the cask, a foot and a half wide, and about two feet high, the door hung with a pair of hinges. This could be opened, and the potatoes taken out with a shovel when needed, or at once emptied into other kits or pails to cool. Or a small loft, or bench, may be built sideways as high as the top of the pot, so that a person can stand upon it and empty the potatoes into other vessels without moving the stand. Indeed this small loft will be needed for filling the potatoes into the stand.

The farmer may use any of the foregoing methods, just as he finds them answer his purpose. I am certain this is the best method of making ready potatoes, either for the family or any kind of stock.

In feeding, great care must be taken not to give them such a quantity of food as may overload their stomachs, which may be in danger of causing them to leave off feeding. If at any time they seem rather to loath their food, and leave a part of it, the only way to remove that, is to lessen the quantity, and now and then change their diet, giving them a few oats, barley, beans, or peas, two or three times a-week, and to be particular in giving them no more than their appetite can devour. The manger, or trough, must be always well washed out before putting in their victuals, as giving them clean food is a leading article to ensure their success in feeding.

If they leave any food in their manger, never offer it to them again, but allow some of the young stock to eat it up, which they will do with a good deal of pleasure. Boiled potatoes, meal, &c. may raise a drought upon them; to remove which, they should frequently have clean water, mixed with a little meal. They ought to have salt to all their boiled food. A little oats for dinner, occasionally, by way of change, will not be amiss.

If the feeder wishes to be profited by this husbandry, he will never allow any of the stock to get poor; for in that state they will consume much more food than is required to keep them in good condition. It is too general a practice to pay no attention to feeding them until they are put up, perhaps in November or December, when with a little pains they might be as fat, and weigh nearly as much, at the time they begin to feed, as at the period when they are fed; which would save two months feeding. Whoever may keep the number of swine formerly mentioned, should have at least ten or twelve acres of potatoes. This would enable them to have old potatoes until the new ones are ready for taking up, and would be of great advantage when keeping so many swine. During the summer, the breeder and feeder of so valuable an animal ought assiduously to purchase wheat bran, barley dust, &c. and keep as much of the refuse of grain as possible. Let them have a little of such articles twice a-day, mixed with water. With those, together with what clover and grass they get in the fields, pigs can make a very good shift until the time for close feeding.

There is no doubt but all this will be attended with some outlay of money, but it must be considered, that it is only the expence of keeping, as there is nothing to pay for the stock. They will handsomely repay all the labour and cost. A part of the crop may be as readily, and with as much advantage, turned into money, in feeding swine, as if sold in the market. Swine are always sold for ready money, and so much rich food being consumed and turned into dung, is certainly a great consideration to a farmer; that dung being worth double the value of any other farm-yard composition. If proper care is taken, a feeding swine should not consume more than six Winchester bushels of oats made into meal. It ought to be shelled before it is ground, the same as for family use, but need not be sifted. It will be understood, that six Winchester bushels of oats made into meal, is only required for those swine intended to be made fat for manufacturing into bacon and hams; porkers do not require such feeding.

The swine, though exceedingly voracious, will feed almost on any thing. In miry and marshy ground (from which they are not averse,) they devour worms, frogs, fern, rush, and sedge-roots. In drier, and in woody countries, they feed on haws, sloes, crabs, mast, chesnuts, acorns, &c. and on this food they will grow fleshy and fat. They are a kind of natural scavengers; will thrive on the trash of an orchard, the outcasts of the kitchen, the sweepings of barns and granaries, the offals of a market, and most richly on the refuse of a dairy. If near the sea, they will search the shores for shell-fish; in the fields, they eat grass; and in great towns, they are supported chiefly by grains. It is evident, that the facility of feeding them everywhere, at a small expence, is a national benefit, more especially in a country where the people are accustomed to eat flesh daily; or, on the other hand, where there is so ready a market for bacon and pork as we have. It is no less observable, that notwithstanding the facility of feeding, and the multitude of swine maintained, they seldom fail of coming to a good market. In no part of Europe is the management of these creatures better understood than in Britain, but more especially of late in Dumfriesshire.

Swine ought to have some hard feeding two or three weeks previous to their being killed, to give firmness to the flesh. This practice ought to be particularly attended to by those who feed at distilleries, on burnt ale and grains, as the fat of pigs thus fed melts almost wholly away in boiling or roasting; pease or beans are excellent for the purpose, and acorns are still better. Where oak plantations are near, they will resort to them in autumn, and there remain until this their favourite food is exhausted. Sir James Colquhoun of Luss, I have been told, was in the habit of sending his pigs to one of the islands of Loch Lomond, where there is an oak plantation, that they might pick up the acorns; which is said to have given a surprising degree of delicacy to the flesh. Those that have woods of this kind, and orchards, ought to allow their pigs liberty to range among the trees, to pick up any shaken fruit, &c.

CHAP. VII.

The time the Breeding Sow ought to be put to the Boar, so as to bring, at proper seasons, two Litters within the year.

EXPERIENCE will teach a farmer who prefers *breeding only*, and sells the offspring, to keep those sows that bring forward the greatest number of pigs and of the best quality, or those that are good nurses, and most careful of their young. When these are met with, let them be preferred while they continue to breed. There is a material difference among breeding sows. When a breeder gets too old, she is apt not to be so careful of her pigs as when young. She will do very well three or four years. After that age, and after having had many pigs, the carcase will not be worth so much as a young sow's by 1s. 6d. per stone; yet they are saleable when well fed. Proper attention ought to be paid to bring the two litters within the year at proper seasons. One litter must be dropped the first week of February, consequently the sows ought to be put to the boar in the beginning of October preceding. If later the pigs would be too young to be fed off next season. The other litter should be in the first week of August; which is late enough, on account of the approaching cold season, though they will make excellent swine to run over the ensuing winter and summer. They will be ready to put up to feed by the 1st of October, and be fat for the market by the end of November, at which time the following February pigs may be put up, and have time enough to feed to perfection before the season is concluded. The feeder may, by this plan, feed two lots in one season. This is a good method, and worth attention.

When a number of breeding sows are kept, it will sometimes happen that part of them may be too late in the season before they farrow, perhaps so late as November, which is the worst season in the year. By the time the pigs are weaned, the days are short, frosty, and cold. There is some difficulty in bringing them through the winter, unless they are particularly attended to. Last season I had fifty-nine pigs farrowed in October, which was occasioned by a neighbouring farmer's boar accidentally getting amongst my sows that were intended to be fed off. I divided them into two lots, gave them plenty of litter in their sleeping apartments, and fed them with boiled potatoes and meal. They never were permitted to go out of the house until the days got longer, and the whole throve remarkably well.

The breeder ought to attend particularly to his sow when the pigs are taken off, that she may take the boar in a week or two after, and not lose the proper season for bringing forth her young. Six or seven weeks is a sufficient time for the pigs to suck, if they are properly attended to. As soon as the pigs are weaned, feed the sow well with good old oats, or meal, and she will take the boar in a few days. The breeder must always sell his pigs when they are taken from their dams, as at that time they have the plumpest look, and will sell for as much money as they will do two months after, unless he has conveniencies to keep them until they are 'shots.' Indeed the August litter will pay very well for running over the winter, and selling at seven or eight months old, as the demand is very much restricted for pigs after the winter commences.

The farmers in the south of England draw large sums of money for swine. From a calculation made upwards of twenty years ago, it appears, that more than 3550 swine, 2515 pigs, were then consumed in London every week. The swine, no doubt, would be made use of in pork, and the pigs for roasting. Many of these farmers keep a certain number of breeding sows; in that case, they are enabled to sell a lot of porkers every two or three weeks, perhaps worth 40s. per head. Wherever they have an opportunity of selling pork at all seasons, they do not think it necessary to make the sows bring their litters at a particular season, as they wish to have a lot of a certain age to go off regularly, at least every month. They make them ready for the market with little expence, only giving them close feeding two or three weeks previous to their being sold. They have very little trouble in selling them, as there are Jobbers continually travelling through the country, purchasing swine of all descriptions; who receive them, and pay the money at the farms. Many of the farmers near London kill their swine at home, and carry them to market themselves. I have often seen a thousand dead carcases of swine in Leadenhall market in one day, mostly brought in by the feeders. Those farmers that live within one hundred miles of London have the advantage of us for marketing their stock, having so regular a consumpt for every article.

CHAP. IX.

Description of the best Conveniencies for Keeping Swine.

I SHALL now give such an explanation of what is necessary for the accommodation and management of swine, as I hope will be the means of promoting peace and tranquillity between them and their owners, and prove to their mutual advantage.

The conveniencies most in use in Annandale, consist of a small hut six feet square, with a court before the door of the same size, where the swine are fed. This method I have tried, but do not approve of it. When the weather is wet, the animals are starved, and their food gets damaged. In frosty weather, it is frozen. There is also another inconvenience—when the keeper puts their food into the troughs, the whole crowd about it, and may abuse it.

My conveniencies at present, which plan I can recommend to the public, are as follow. A house thirty by fifteen feet, made with four doors, which open outwards (the litter in the inside would prevent their opening in the usual way), and three partition walls through the house, viz. a wall betwixt each door, which divides the house into four apartments. The two in the middle I adapt for eating rooms, and the others for sleeping apartments; having an inner door betwixt each eating and sleeping apartment. By this plan the keeper gets their eating rooms swept out, their mangers cleaned, and their food put in, without any incumbrance from the swine, as they are all at rest. When this is done, unbolt the door and ring the dinner bell. I might add, that a division wall would be proper through each sleeping apartment, with an open door in it. The apartment next to the backside of the house is to hold the litter, where the family sleeps. The apartment next the front of the building, which they walk through to their eating room, will serve as their necessary, as they are so cleanly that they will not defile their bed. The latter place may be the smaller of the two.

The most convenient manger for their food is as follows. Let it be as long as the house is wide, and fixed against the middle wall. It may be similar to that of a horse manger, only not so deep, and must be divided into twelve divisions by partition boards. Let each of these boards be cut three feet in length, or height, and a little broader than the manger is wide; fix them to its bottom, and nail a board, or thin plank, over the top of these fixtures. This method prevents the swine getting into the manger and destroying their food. In this way the number before mentioned will feed as quietly, and thrive as well, as if there were only two or three together.

Before every meal the manger ought to be washed, and the room clean swept out. A little clean litter, every night, will make the family comfortable. I have at present thirty young swine that sleep and eat together, and I never remember of having any that throve so well. Each of the abovementioned sleeping and eating rooms, can be divided into two, in case they are required for sows and pigs, and the partitions again removed at pleasure.

I would recommend a still further improvement to those who may have their conveniencies yet to build, viz. five apartments in place of four. The middle apartment to have a fixed boiler in it to make ready their victuals, with a chest to hold their meal, bran, barley dust, &c.; with an eating house at each end of the boiling house, and the sleeping apartments adjoining.

Their sleeping apartments should be dark, close places, which is a great advantage for their fattening. When they are allowed to ramble up and down, even after a good meal, the food has not the same effect. Every apartment ought to be paved, so that they can be kept clean and dry. There should be a square, or yard, formed by the huts or other buildings for sows; and that square likewise ought to be paved. If about an acre of land could be conveniently enclosed with a stone dike, adjoining their houses, it would be of great use for the store pigs and sows.

I have had, in the course of more than thirty years' practice, many different plans of swine-houses, but I have found what I have now described greatly superior to any of them. When this plan is completed, it will enable the breeder and feeder to manage these animals properly, which, I hope, will be a means of removing a part of the prejudice entertained against them, and of rendering them a more valuable sort of live stock than otherwise, if ill accommodated and inattentively managed.

CHAP. X.

Proper Directions for Killing, Cutting up, and Curing and Drying the Flitches and Hams, for the London Market.

It would be superfluous to say much upon the method of killing swine, as it is so generally known. I should recommend at first to give the animal a knock on the head, to make it lie quiet, and then lay it flat on a level piece of ground, or board, upon its right side. Let the butcher then stand straight before it, with the edge of his knife towards its belly, and thrust it straight forward before the brisket, towards the heart. The back of his knife will nearly touch the backbone of the sow, and he must level his knife so as not to get it into the first rib. After the animal is dead, and laid upon some board, or table, pour boiling water over it out of a tea-kettle, or something similar, and immediately set to work with a scraper, or knife, which will clean off the hair, &c. and so go on pouring and shaving, until the carcase is quite clean.

In many countries they singe off the hair with burning straw; but the former method seems to be most generally practised in this country, and, I think, answers the purpose best.

After the carcase has hung all night, lay it upon a strong table, or bench, upon its back, cut off the head close by the ears, and cut the hinder feet so far below the hough as will not disfigure the hams, and have plenty of room to hang them by. Let the curer then take a cleaving knife, and if he chuses a hand mallet, and divide the carcase up the middle of the backbone, laying it in two equal halves. Then cut the ham from the side by the second joint of the backbone, which will appear on dividing the carcase; then dress the ham by paring a little off the flank, or skinny part, so as to shape it with a half round point, clearing off any top fat that may appear. He will next take off the sharp edge along the backbone, with his knife and mallet, and slice off the first rib next the shoulder, where he will perceive a bloody vein, which he must take out, for if it is left in that part is apt to spoil. The corners must be squared off where the ham was cut out.

In killing a number of swine, what sides you may have dressed the first day lay upon some flags, or boards, piling them up across each other, and giving each flitch a powdering of saltpetre, and then covering it with salt. Proceed in the same manner with the hams, by themselves, and do not omit giving them a little saltpetre, as it opens the pores of the flesh to receive the salt, and besides gives the ham a pleasant flavour, and makes it more juicy.

Let them lie in this state about a week, then turn those on the top undermost, giving them a fresh salting. After lying two or three weeks longer, they may be hung up to dry in some chimney, or smokehouse. Or, if the curer chuses, he may turn them over again, without giving them any more salt; in which state they may lie for a month or two without catching any harm, until he has convenience for drying them.

I practised for many years the custom of carting my flitches and hams through the country to farmhouses, and used to hang them in their chimneys and other parts of the house to dry, some seasons to the amount of 500 carcases. This plan I soon found was attended with a number of inconveniencies, having to take along with the bacon pieces of timber, to fix up in the different houses, for the purpose of hanging the flitches and hams. For several days after they were hung up, they poured down salt and brine upon the women's caps, and now and then a ham would fall down and break a spinning wheel, or knock down some of the children; which obliged me to purchase a few ribbons, tobacco, &c. to make up peace. But there was a still greater disadvantage attending this mode; the bacon was obliged to hang until an order came for it to be sent off, which being at the end of two or three months, and often longer, the meat was overdried in most places, and consequently lost a good deal of weight. This method is practised at this day in Dumfriesshire. People in general are so partial to old customs, that it is nearly impossible to remove them.

About twenty years ago, to prevent these disadvantages, I contrived a small smoke-house, of a very simple construction. It is about twelve feet square, and the walls about seven feet high. One of these parts requires six joists across, one close to each wall, the other four laid asunder at proper distances. To receive five rows of flitches, they must be laid on the top of the wall. A piece of wood strong enough to bear the weight of one flitch of bacon, must be fixed across the belly end of the flitch by two strings, as the neck end must hang downwards. The piece of wood must be longer than the flitch is wide, so that each end may rest upon a beam. They may be put so near to each other as not to touch. The width of it will hold 24 flitches in a row, and there will be five rows, which will contain 120 flitches. As many hams may be hung at the same time above the flitches, contrived in the best manner one can. The lower end of the flitches will be within $2\frac{1}{2}$ or 3 feet of the floor, which must be covered five or six inches thick with saw dust, which must be kindled at two different sides. It will burn, but not cause any flame to injure the bacon. The door must be kept close, and the hut must have a small hole in the roof, so that part of the smoke may ascend. That lot of bacon and hams will be ready to pack up in a hogshead to send off in eight or ten days, or a little longer, if required, with very little loss of weight. After the bacon is salted, it may lie in the salt-house as described until an order is received, then immediately hang it up to dry.

I found the smoke-house to be a great saving, not only in the expence and trouble of employing men to cart and hang it through the country, but it did not lose nearly so much weight by this process.

It may be remarked, that whatever is shipped for the London market, or any other, both bacon and hams, must be knocked hard, and packed into a sugar hogshead, or something similar, to hold about ten hundred weight. Bacon can only be cured from the middle of September until the middle of April.

After the introduction of the smoke-house into the country, the manufacturers continued the practice of carting the bacon through the country, and hanging it in farm-houses. Many are averse to adopt any new system, even though they are convinced of its utility. One of our principal dealers would not make use of a smoke-house, I suppose because he was not the propagator of it; but introduced a stove, which he placed in an old cow-house, and hung his bacon and hams all around it. In consequence of this, a great part of his bacon and hams became sour and ill-tasted; not being dried with wood or peat-smoke, they had no flavour; so that most of them were spoiled, and not marketable: and I understand he lost by this some hundred pounds.

Juniper wood gives bacon the best flavour; sawdust answers the purpose very well, and is most easily got. The bacon and hams in Jersey and Guernsey are delicious, on account of their pigs being mostly fed with fruit, and likewise the inhabitants burn seaware in their houses for fuel; the hams being dried in the chimneys, gives it an agreeable flavour.

In Berkshire they have a little close room in the garret, where they smoke their bacon and hams; the place is connected with the kitchen chimney, where it is filled with smoke as it ascends; it is entirely close, and no entrance to it but up the chimney, where the people ascend to bring out their bacon, flesh, or ham. This is an effectual method to prevent the bacon being struck with the maggot-fly, which is a great nuisance to meat hanging in open houses.

CHAP. XI.

A Description of Pickling and Kitting Pork, &c.

This branch of manufacturing pork has been for some time established in the neighbourhood of Berwick. The carcase is cut in pieces, and packed in kits made for the purpose, which contain some of them 1 cwt., others 1½ and 2 cwt. Salt is dissolved in cold water, so as to be strong enough to swim an egg; then poured into the kit upon the pork until it cover it. When the end of the kit is fixed in, the article is ready for being sent to market; and, I understand, there is as great a demand for it as for bacon and hams.

In Dumfriesshire, Galloway, and Cumberland, this mode of preparing pork is yet unknown; but it would certainly be a great advantage, were it to become general in these counties, where so much pork is made into bacon. To my knowledge, near the end of the curing season, many young pigs half fed, are exposed in the market for sale, not fit to be made into bacon, which would answer the purpose of pickling. In this case, all the old and well fed swine might be manufactured in the former way, the inferior sort in the latter. Thin swine lose more weight in drying than well fed ones, and do not take the market so readily, nor yet bring so good a price. I should recommend both practices to be carried on together, as the variety of stock affords different kinds of pork, suitable for the two ways of preparing it. Swine for pickling do not require such high feeding as the other; they would not answer the purpose if they were too fat.

A farmer having convenience for keeping five breeding sows, may have for pork eighty pigs every year. The February litters would be ready for pickling the latter end of October, just as taken from the stubbles, without being put up to feed with meal. It may be necessary, however, one month previous to their being killed, to let them have plenty of raw potatoes thrown among them every morning and evening. As I have formerly stated, always let their bellies be full of something. If that is attended to, when eight or nine months old, they should weigh nine stone, and sell for 6s. per stone (which is below an average price for some years back), and so would bring £.216. By this plan the farmer has only 40 pigs on hand at once. By the time the August litters are taken from their dams, the February ones will be nearly ready for slaughtering. The former lot will be fit to kill at the end of March, at which time the next February litters will be weaned. Five sows must be kept from every February litter for breeders. The five sows will in this case be fed after having had two litters of pigs, and will at that age weigh twenty-four stone each, which, at 6s. per stone, is £.36, which sum, in addition to the produce of their offspring, makes £.252.

The number of swine may be proportioned to the size of the farm. In my opinion, this method will bring in as much money to the farmer, and not nearly cost so much in feeding as the other. It will be observed, that when the pigs are arrived at the weight of nine stone, they are heavy enough for the purpose of pickling, and at that period they are ready, if the farmer chuses, to be put up to feed for bacon. It will at least take three months before they can be fed to the weight of eighteen stone; and they will consume one stone of meal each besides potatoes, for every stone of pork that they increase, which is little more than $1\frac{1}{2}$ lb. of meal each per day. There will not be so much clear profit from the increase of the last nine stones as from the first, as the quality or expence of the food makes a material difference, besides firing and attendance, &c. But in districts where the pickling process is not known, those who keep swine are obliged to feed them for bacon. There must likewise arise another advantage to the person who pickles pork, which, in the present state of the market, is not possessed by the preparer of bacon, in respect the former can put it up in kits one week and ship it the next, at which time he can draw upon his correspondent for the amount, while the latter lies out of his money from three to nine months, besides incurring a deal of more labour and risque.

With the hints mentioned and a little practice, the farmer will soon discover which method is most profitable for himself to adopt in his own situation. When the pickling process becomes general it will be much in favour of the cottager, as he will then be able more easily to keep two swine than one at present, when frequently they are not able to complete the fattening of their pigs for want of meal, and are obliged to sell them much under market price. Husbandmen and cottagers should prefer either of the methods, and should prepare either pork or bacon as may suit best their convenience, and the markets they have: It cannot well be expected that the same mode of management will answer with advantage in every place. It is enough to suggest, that pickling pork wants only a beginning here to make it spread farther, as bacon curing does in other districts.

Since the publication of this treatise, I have been informed that the trade of pickling and kitting pork for the London market, has been carried on for a number of years by several companies in Inverness,

Cromartie, &c. This trade has been considerably increased since the inhabitants of these districts had an opportunity of perusing what I have written on the subject. The district is well calculated for carrying on the trade, as they have plenty of pigs of a proper size, and as they are upon the spot from which vessels sail to London.

CHAP. XII.

Of the Making of Brawn. Receipt for Preventing Bacon from turning Rusty. How to Cure the Measles, and the Rupture, in Pigs.

BRAWN is the flesh of a boar soused, or pickled; for which end the boar should be old, because the older he is the brawn will be the more horny. It is the flitches only, without the ham or shoulder, that are made use of for this purpose. Brawn is prepared in the following manner: After the boar is killed, the flesh is to be sliced off the backbone and ribs, and afterwards sprinkled with salt; it must then be laid in a tray till the blood be drained from it; give it a little more salt, and let it be rolled up as hard as possible.

The length of the collar of brawn should be as much as one side of the boar will bear, so that when rolled up it will be nine or ten inches in diameter. After being rolled up, it must be boiled in a copper till it is so tender that you can run a straw through it; then set it past till it is thoroughly cold, when it must be put into the following pickle. To every gallon of water put a handful or two of salt, and the same quantity of wheat bran; boil them together, drain the bran as clear as you can from the liquor, and when it is quite cold put the brawn into it.

In order to prevent bacon from becoming rusty, or moulded, after it has been salted about a fortnight, put it in a box, or tub, the bottom of which must be covered with sweet hay; and between every piece of bacon put a layer of hay. This will prevent it from rusting, and keep it above twelve months as good as at the first day: this will also keep your meat in a pickled state without its being hung to dry. It will be necessary to keep the box or kit shut up, to prevent mice or rats getting in.

It sometimes happens, though seldom, that swine have the measles; while they are in this state their flesh is very unwholesome food. This disorder is not easily discovered while the animal is alive, and can only be known by its not thriving or fattening as the others. After the animal is killed and cut up, its fat is full of little kernels, about the size of the roe or eggs of a salmon. When this is the case, put into the food of each hog, once or twice a-week, as much crude pounded antimony as will lie on a shilling. This is very proper for any feeding swine, even though they have no disorder. A small quantity of the flower of brimstone also, may be given among their food when they are not thriving, which will be found of great service to them. But the best method of preventing disorders in swine, is to keep their sties perfectly clean and dry, and to allow them air, exercise, and plenty of clean straw.

Where a number of swine are bred, it will frequently happen that some of the pigs will have what is called a 'rupture;' *i. e.* a hole broken in the rim of the belly, where part of the guts comes out and lodges betwixt the rim of the belly and the skin, having an appearance similar to a swelling in the cod. The male pigs are more liable to this disorder than the females. I never found much difficulty in curing this disorder, by the following means.

Geld the pig affected, and cause it be held up with its head downwards; flay back the skin from the swollen place, and from the situation in which the pig is held, the guts will naturally return to their proper place. Sew up the hole with a needle, which must have a square point, and also a bend in it, as the disease often happens between the hinder legs, where a straight needle cannot be used. After this is done, replace the skin that was flayed back and sew it up, when the operation is finished. The pig should not have much food for a few days after the operation, until the wound begins to heal. I might have described the method of spaying female pigs; but as it cannot be well understood without one sees it performed, I judge it better to decline it.

CHAP. XIII.

On the Supply of Fresh Pork throughout the Year to Private Families, at a small expence.

To gentlemen and farmers who are obliged to keep large families, may be recommended the following plan of supplying themselves with fresh meat, which is one much practised in the south of England.

Purchase a proper sow, or sow pig, of the real Chinese, or thick, round, black breed. That single sow will supply a family throughout the year with little expence. She will bring up from eighteen to twenty pigs yearly[*], and that sort may be made fat for pork when they are from four to six months old. Allow the same sow to breed so long as she proves a kind mother and good nurse; but when she turns the contrary, feed her off and put a young one in her place.

The farmer, or some of his neighbours, may have a boar of the same breed, or if he be of a larger sort, it may do for one cross, but the sow should always be of the sort described. Build a small house, such as will admit two of these pigs to feed in, and let it be as near the back kitchen as possible, for servants dislike to go to a distance to give them any

---

[*] A black Chinese sow, belonging to Lady Robert Manners, littered 26 pigs, all alive and healthy, at her Ladyship's late Manor-house, Sutton Surrey.

trifling offals from the table or kitchen. A pig may be easily fed for pork on the refuse of a farm house, and the expence never felt. It may be proper to have always two in the sty at a time, and one may be killed every three weeks*. They will be ready for slaughter about six weeks after they are shut up. When the oldest residenter is killed, put up another to the one left, and so on in rotation.

In this manner a few swine are sufficient to supply plentifully a table every day at a small expence. The Scots in general, particularly the lower classes, are not fond of pork; but the cause of this dislike may arise from its being so scarce an article. Besides, swine that are reared in Scotland are kept to a great age, fed to a great weight, and made into bacon, which is much stronger to the taste. Proper pork is a delicious dish, either fresh or pickled. The

---

* It may be argued that June, July, and August, are not proper months for killing pork pigs, as they cannot then be properly pickled. The pickling, however, may be done even in these months, in a cold cellar, sufficient to serve any length of time that such pork would require to be kept. In the family of an ordinary farmer, if there be a good housewife, she may easily render it unnecessary to kill pigs for three months, by having seven or eight broods of young ducks and chickens, which may afford three or four brace every week for the family table during that time. As it is a plentiful season for milk and butter, the servants can easily be supplied from the produce of the cow; and as every farmer that feeds swine has always a few flitches hung up, let them have a slice of it with beans, by way of a change.

housewife might cure some of these little hams in the Westphalia form, at present so much in vogue. The method of doing this is, to sprinkle the ham over with salt, and let it lie for twenty-four hours; then take out any blood that may be in it and wipe it dry. Mix a quantity of brown sugar, half a pint of bay salt, and three pints of common salt (proportioned to what quantity of hams are used), well stirred together in an iron pan over the fire, till it is moderately hot, and let the ham lie three weeks in this pickle, frequently turning it; after that dry it in a chimney.

CHAP. XIV.

A Statement of the number of Swine fed in Annandale yearly, their average weight, and value. Importance of the Pork Trade to Bankers.

HAVING been at a good deal of pains, by corresponding annually with dealers and intelligent farmers, almost in every parish in Annandale, and having myself formerly purchased the most part of the swine fed in the parishes adjoining to my residence, I reckon the following an accurate statement of the number, weight, and amount of value, of the pork produced in one season, in Annandale, consisting of above twenty parishes.

It appears, from the strictest investigation, that 10,000 swine are fed annually, weighing at an average 14 stone each, making 140,000 stone, at 7s. per stone. But say every sow is worth £.5,) head and feet being left at home,) is £.50,000. A sow will have one pound of lard for every stone of pork, say at 6d. per pound, is £.3,500, making a total amount of £.53,500.

There will be about the same number of swine fed in the remaining part of Dumfriesshire and Galloway. It is mostly all sent to the London market, and pays a duty when entering England, of 3d per stone, which amounts to £.1750. The one half of the sum charged is more than sufficient to pay the difference of the duty on salt. Such an exaction would seem to be an improper advantage taken of this country, through our own neglect. Our landed gentlemen ought to inquire into this matter, which operates as a considerable discouragement to a trade which is so beneficial to the public, and without which the landlord's rents and taxes could not be readily paid.

There is no trade carried on in Scotland so beneficial to the banking business as that of bacon and pork. The notes of the banks being paid by traders to farmers and cottagers in sums of from £.5 to £.30; this money is partly laid up for three or four months for payment of rents, and partly laid out upon necessaries for the family, some having oats to buy, some a horse, &c. thus affording a wide circulation for the notes, whereas what money is issued to drovers, &c.

the notes are paid in large sums, and will be back upon the bank in a few days or weeks. The manufacturers of pork and bacon send it all to the London market, and receive in return Bank of England paper, or drafts upon the London merchants, which all come through the hands of the Scots bankers.

To my certain knowledge, the two branches of the Edinburgh banks in Dumfriesshire, with the Castle Douglas bank, for some years past, have issued upwards of £.100,000 every season, to be laid out in pork. The bankers, therefore, should encourage the establishment of such a beneficial trade. The amount of the value of swine annually fed in Dumfriesshire is, however, considerably more than this. Pork has been selling this season (1814) in Dumfriesshire, at from 10s. to 10s. 6d. per stone, of 14 lb., the carcase, head and feet off. I never knew the wholesale price of pork above 9s. per stone before this period; and this holds out an additional inducement, were any wanting, for this valuable stock being increased. No pigs can be got in Dumfriesshire or Cumberland under 30s. per head, just from the sow.

Although the number of swine reared in Dumfriesshire may appear great, yet the people employed in the curing of bacon cannot be supplied with fat swine sufficient to answer their demand. It appears to me, that the farmers in many places of Scotland are deferring increasing their stock of swine until they see a market established for them: On the other hand, those that would embark in the trade of curing seem to be waiting until the farmers increase their stock of pigs, so that it might be worth their while to make a beginning.

CHAP. XV.

On the Breeding of Swine in the Highlands and Western Isles of Scotland.

Swine do not appear to be used in the Western Isles. Dr Johnson, in his Tour to the Hebrides, in 1773, says, that in Sky the people held pork and bacon in abhorrence, and accordingly he never saw a hog in the Hebrides but at Dunvegan. But I should suppose that these islanders, by this time, have laid aside their prejudices, and found out the benefit of rearing such a profitable stock, if they have not yet come to the knowledge of bringing up and marketing swine. I would recommend it to the inhabitants of these islands (many of whom I have been long acquainted with), that they should immediately introduce this stock into their respective abodes. They should be of the true Highland breed, such as are common in Caithness and Sutherland. No where do I know a country so well adapted for rearing swine as the Western Isles, where there are large tracts of pasture lands, in which swine will feed all the summer months among their cattle and sheep, and will resort to the shores and pick up sea-weed and shell fish, which are excellent food for them. It would be advisable to erect a few huts in different places where they pasture, so that the pigs may lodge all night, or in the time of a heavy rain; a few rushes may be thrown into their huts for them to lie on.

Were any gentleman or farmer that has any extent of land in hand to keep four or five breeding sows and a boar, he could send to market every year sixty or seventy young swine, at the age from six to twelve months. It is not meant that these breeders are to make their swine fat, but only prepare them to be fed off in the corn districts. As many potatoes should be grown as possible, for subsisting them during the winter. Droves of swine might be sent from these islands from the month of August to the end of November, the Falkirk and Doune fairs being very proper times for the disposal of such stock. Jobbers from the south, even from Dumfriesshire, would be glad to meet with droves of swine at these markets, and they would meet with as ready a sale as the cattle and sheep brought from these isles. It is very common for droves of swine to be driven many miles from the interior of Ireland, to be shipped for Liverpool, Bristol, and London. I have seen upwards of a hundred in a drove shown at Edinburgh, from Ireland, all conveyed to these places by Jobbers. Now they might just as easily be brought from the islands of Scotland. The inhabitants of these isles have long sent their cattle over to markets in the mainland, even at a time when they did not bring more than from 30s. to 40s. per head; such now will bring £.8. In those days swine were not worth sending to market. I have purchased droves of them twelve months old at 5s. per head; such now would bring 20s. The present high price of every article of food, and the cheapness with which swine can be reared in these districts, are certainly great inducements for breeding them. It would also be convenient for those who live at a distance from butcher markets, and cannot be supplied with fresh meat, if they were keeping a number of swine; they might kill one every month, which would nearly keep the family in fresh meat; they might also feed a few of them fat, which they could hang and dry for family use.

I have attended swine fairs at Ovingham, in Northumberland, in the end of harvest, where I have seen nearly a hundred farmers, selling their young swine, about six or eight months old, then called shots, perhaps from 10 to 20 in the hand of the breeder. The total amount of pigs shown in that fair might be from six to seven hundred, almost all of them bought up by Jobbers, the greater part of which were driven into Dumfriesshire, some to the distance of more than a hundred miles, and exposed for sale in fairs and markets, and bought up by farmers for feeding off that season.

I would recommend the same plan to farmers in the counties of Argyle, Inverness, and the northern part of Perthshire, which are all well calculated for the rearing of swine, and are even in this respect superior to the Western Isles, being much more conveniently situated for marketing their stock. Large herds of Highland swine were formerly kept in some parts of these districts, which are now much out of use. The landholders and farmers in these places would do well to encourage the breeding of swine as formerly. They would find a double, nay a triple advantage, by keeping that stock at the present day, as they would bring four times the price they did twenty years ago. The true Highland breed will still be preferable to any other, in high districts, as they are more hardy, and will endure the wet and cold that they will be exposed to in the glens much better than any other sort, and will besides feed to a proper weight for porkers.

The low corn counties around, viz. Angus, Perth, and Fife, would feed more than the Highland counties could produce. In many places where much corn is grown, there is no convenience for keeping swine until the end of harvest; the farmers in these districts would then gladly purchase ten or a dozen swine to put upon their stubbles. Very little additional food would be required for them till they were ready to kill for porkers. Many farmers near Perth, in Stratheam, and the Carse of Gowrie, &c. might easily keep from twenty to thirty of these small swine after harvest to pick up what grain is left in the fields. Give them a few raw potatoes once a-day, and about two weeks or so before they are to be slaughtered let them have a few pease, or any kind of grain twice a-day, to sadden their flesh. They might be all sold off by the end of November. Suppose the farmer were to receive from 20s. to 30s. per head for the short time he would have to keep these animals, he would be well paid for his trouble, and the dung they would produce would nearly pay all the expence. It is not meant that swine of this descrption are to be fed in the way they do in Dumfriesshire, where they are fed at a high expence, with meal, to be manufactured into bacon and ham, and made to weigh from 14 to 24 stone each. Let the farmer begin and keep six or ten at first; he need not doubt of finding a market for them, as I know of people in that neighbourhood who are only waiting the opportunity of being supplied with that stock, in order to commence the trade of kitting pork.

From what we observed when treating on the kitting of pork, the reader will be apprised that the people in the counties of Inverness and Ross, are carrying on that trade with great spirit; several of the traders go as far as Sutherland and Caithness, to purchase droves of Highland pigs, which they drive to Cromarty and its neighbourhood, and slaughter, and kit up the pork for the London market. I trust the period is not far distant, when the manufacturing of pork into bacon will commence in these northern counties, to the great benefit of the inhabitants. The people in the Orkney Islands possess the same advantages for prosecuting this trade, and have water carriage in a manner at their doors.

If there were plenty of swine fed, no doubt markets would soon be established in the different districts, and the swine bought up and manufactured, the oldest and fattest into bacon and hams, and the smaller and half fed into pork for pickling.

CHAP. XVI.

The advantages the Farmers along the Frith of Forth, and the neighbouring Counties, have over those of Dumfriesshire, for Breeding and Feeding Swine, and carrying on the Trade of Curing Bacon.

The whole of the Lothians, that fertile valley on each side of the Forth, as far as Stirling, the coast of Fife, and the Carse of Gowrie, as far as Perth, are all corn countries, which are the only places where this stock can be kept with advantage. There are sometimes late harvests, when part of the crop is damaged by bad weather, which would be very good food for swine. The fields which beans, peas, and other corns, are taken from, have always something left and lost which these animals might gather up.

The districts mentioned have many advantages. They have the salt for curing at a much lower rate, being made in their neighbourhood, while the people in the south of Scotland have it all to purchase at

Prestonpans, or the coast of Fife, and bring it through the canal and round the west of Scotland, at a heavy expence. It is now selling here at 22s. per cwt. They can ship their bacon in a manner at their very door, and have opportunities to London two or three times a week, with smacks which sail from Leith; while the Dumfriesshire or Galloway traders have land carriage to pay to Newcastle, from which they are distant eighty, a hundred, and some of them even one hundred and forty miles.

All the district from St Andrews by Kinross and Stirling, to Glasgow, and even through the west of Scotland, is well adapted for both breeding and feeding of swine, as in that extensive range of country they have both pasture and arable land, where they may feed their swine through the summer, and land where they may raise plenty of potatoes to support them through the winter.

It may be said they have no inclosures to confine them, and a number of swine would destroy their crop? To this I answer, put a herd to them, as they do in other countries, they will pay as well for herding as your cattle and sheep Let them be put into a house all night with plenty of litter, the dung will do more than pay the expence of a herd.

I would recommend it to the farmers of the above boundaries to keep a larger breed than the Highlanders, as their climate is milder, and their keeping better. The kind that will weigh about eight stone, 14 lb. to the stone, or eight months old, and which would feed to 15 stone at twelve or fifteen months old. Swine of this sort will soon rush up to 8 stone. Those now common in the Lothians and Fifeshire, will not get to 8 stone at a year old, and when at the age of a pork swine will not exceed 4 or 5 stone. This kind will not pay the farmer: too small a breed of pigs is not profitable to the feeder. It is most likely that pork pigs will be more inquired after in the above districts than heavy swine; indeed they will pay the feeder most money, and with less expence and trouble. I would recommend the species that is small boned, and as broad over the back as they can possibly be got. Breeders will have it in their power to select a preferable breed by paying attention and picking up a young pig of the above description where-ever they shall chance to meet with it; they should not allow prejudice to hurt their interest, which is too often the case. Be very particular, if you have a well shaped sow, to put her to a boar as near her own make a possible, if you should send her twenty miles. Those who are within reach of Cameron Bridge, near Edinburgh, will find a boar of the right sort with Mr Proud, the miller; and also with Mr Todd, farmer, near Blackness, who have swine of the proper breed.

How was the most valuable stock of horses, cattle, and sheep, brought to such perfection by the late celebrated breeders Mr Bakewell and Mr Culley, but by paying particular attention to picking up a well shaped animal wherever they could meet with it, and always allowing the most proper sort to breed together? By persevering in this plan all the different kinds were brought to great perfection. Swine is of all domestic animals that which is soonest brought to perfection, as they come to maturity in the course of twelve months, and if they should not please a change is soon effected, and the experiment of endeavouring to get into the best breed is not expensive.

The breeders in these districts have an advantage over those in Dumfriesshire, as formerly mentioned, as they will save a handsome profit (viz. four or five shillings per hundred weight) which the latter pay for land carriage. As the latter, however, notwithstanding this, still find it an advantageous trade, as is evident from the spirit and extent to which they push it, I must again recommend the practice to the farmers in the west of Scotland, especially in those districts where they make so much cheese.

The farmers in the west country, viz. in Renfrewshire, Ayrshire, &c. know very little of the advantages of keeping swine. A friend of mine told me, that when he was in Ayrshire, lately, he had occasion to be at the house of a considerable farmer, who kept twenty milch cows for cheese making. He inquired at the farmer why he did not keep some swine to consume his whey, &c. In reply, his friend answered, that he might venture upon *one*; but he did not know where it could be procured, as there was only one little black sow that he knew of, in that neighbourhood. My friend said he might easily keep *two* pigs. Had I seen the gentleman, I should have advised him to keep a pig for every cow. For my part, I do not know what use they make of their whey, &c. as they bring up few calves. I would strongly recommend the swine husbandry to be practised in those districts, where they make so much cheese, which produces so much offals for pigs.

Every cow will produce from six to eight pints of whey every day, that of itself, with grass, is sufficient to keep the pig in good condition. When the season for cheesing is over, let them go in the fields after harvest, and pick up what is left; then let them have potatoes and grain of any sort to sadden their flesh and make them ready for the merchant. At the same time the farmer should have some wheat bran, barley dust, &c. and put a handful in to thicken their whey; that would not be lost, as it would add to their weight. He that keeps twenty cows for cheesing, by this calculation, keeps twenty swine. Suppose the feeder was to lay out ten or fifteen pounds in wheat bran, barley dust, &c. that number of pigs would pay him from forty to sixty pounds yearly clear money, besides the valuable dung, where at present he does not make as many shillings. Enough has been said to lead people in this district to make a trial of the swine husbandry, and if once they had their eyes opened to its advantages they would need no more encouragement or direction, as its profits would be sufficient inducement for them to persevere.

One great drawback on this extensive trade, is the paying such a heavy duty for sending bacon from

Scotland into England, no less a sum than 2s. per cwt. being charged as the supposed difference betwixt the English and Scots salt duty This is an absurd calculation! the bacon curers ought in time to have acquainted landed gentlemen, and the representatives of boroughs, with that unfair charge, and I make no doubt they would have had it in their power to prevent such an act passing.

About twenty years ago I had five cart loads of bacon stopped at Carlisle, and a duty of 1s. per cwt. demanded by the customhouse. As it was the first time that any duty was asked, I signified a wish to know on what account this duty was charged. In reply, I was informed that it was the difference between the English and Scots salt duty, the bacon being cured with Scots salt, and sent into England for sale. I observed that their calculation was too high, and that seven pounds of salt would cure one cwt. of bacon. On making a statement on the difference of the duty, I told them sixpence per cwt. was as much as could be charged. The bacon was detained, however, until the customhouse had an answer from the board. The answer was to charge sixpence per cwt. which I paid and received my goods. This duty of sixpence per cwt. was charged for a long time after this, and before any bacon was allowed to enter into England, the trader was obliged to go to the customhouse at Carlisle and take out a permit for the quantity intended to be delivered.

For some time past a duty of 2s. per cwt. has been

charged; upon what ground this additional charge is made, I do not know. Since the duty of 2s. per cwt. has been charged, the taking out of permits has been abolished, and an excise officer attends at Longtown, where he sees all the bacon weighed, and receives the duty. All the bacon passes this place in its way to Newcastle, from whence it is mostly shipped for London. The public will see the necessity of our members of parliament using their endeavours to get this overcharge of duty taken off: It may be proper that a duty should be paid, but not a treble one. I am fully convinced this matter only requires a little discussion to get it rectified. The gentlemen in the Lothians would probably look better to their own interest than allow any duty to be charged but what is right.

After making strict inquiry concerning the difference of the Scotch and English salt duties, I find that Scotch salt pays 6s., and the English 15s. per bushel; 56 lbs., or half a hundred weight, of salt, is sufficient to cure eight hundred weight of bacon, consequently the difference on the duty of salt that is used for curing that quantity is exactly 9s. At this time the customhouse of Carlisle charges 2s. per cwt., which makes 16s.; now, in place of 2s. being taken, 1s. $1\frac{1}{7}$ is the exact difference per cwt. that ought to be charged. As this imposition is very prejudicial to the trade, it is hoped some of the members of parliament for Scotland will inquire into it and get it amended. The people also, on the borders of Scotland, are not exempted from the duty, even although they cure their bacon with English salt; and though they proposed to bring receipts from the merchants of whom the salt was purchased, to the full amount of the bacon cured, and offered to make oath that the bacon was all cured with English salt. The bacon manufacturers might be allowed salt duty free, as that trade is as beneficial to the community at large as the fishing trade.

There are upwards of fifty tons of hams and flitches sent from Annandale annually to Edinburgh and Leith: a carriage of eighty or ninety miles, which certainly ought to be another consideration for the people of the Lothians to supply their own country with this article. I hope all these advantages, when considered, will in some measure convince them how much the Lothians and neighbouring counties are adapted for the carrying on of this valuable trade.

CHAP. XIV.

Practical Remarks in addition to those already offered.

There is little risk attending any of the plans laid down, as there is very little money sunk in stock. Twenty guineas will nearly purchase all the necessary stock, and in ten or twelve months the farmer will reap the benefit. It is indeed hardly possible there can be any loss, for the whole stock could be immediately turned into ready money. If any person, where the trade is unknown, should wish to make a trial, he need entertain no doubt of finding a market for his pork. Although the sum is large that is laid out upon this article every season, yet the purchasers would buy as much more were it to be gotten. I frequently see the merchants in that trade striving who shall get the first offer of any lot intended for sale. Several people have gone from this country to Ireland, bought pork, and cured it, sent it thence to Annan by sea, forwarded it to Newcastle with carriers, and then again shipped it for London.

Our north country neighbours ought to consider the advantages they have for pursuing this branch of business, which is certainly well worth their attention, especially as pork (the whole carcase except the head and feet) has, for several seasons back, brought from 7s. to 9s. per stone, and was so high this year (1814) as from 9s. to 10s. The young stock sold equally high; a small pig of six weeks old, that one might have put in one's pocket, gave 20s.; the best sort, at eight weeks old, from 25s. to 35s. Indeed some of my neighbours, in May last, sold all their pigs so high as 36s. and could not supply half their customers. Those who peruse the public prints will observe the unlimited demand in all quarters for store pigs.

About twenty years ago I attended Edinburgh market for a few weeks, and picked up all the hams I could get among the butchers, as nobody else had any for sale. I bought them considerably lower than

I could then do in Dumfriesshire, and had a saving in the salt, as the women brought it from the salt works on their backs, and laid it down in the curing house. But not finding a sufficiency of hams to make it worth my notice, I relinquished the business, and sold my hams in the salt to Mr Robert Johnston, merchant in Edinburgh, and left him to manage the process of drying, after giving him the proper information as to smoaking them with saw-dust, &c.

It is rather remarkable that there is no person north from Langholm in the bacon trade. Many carcases of swine from Jedburgh, Selkirk, and Hawick, are sent to the Langholm market, which are generally bought by people from the neighbourhood of Longtown, cured and sent to Newcastle to be shipped for the London market. If any person in the districts mentioned were to begin the trade there would be a great saving in land carriage, as Berwick is as good a port to ship at as Newcastle. In a short time, perhaps, all these advantages will be found out and fully appreciated.

CHAP. XIX.

Of the Prejudices so unjustly entertained against this Species of Stock.

I HAVE only now to endeavour to do away the prejudices which exist with many against swine. In ancient Egypt the hog is said to have been esteemed so very unclean, that any Egyptian who had touched it by accident plunged himself into the river. The persons employed in feeding and rearing this animal, are said to have been detested by the rest of the community; their sons and daughters married among themselves, as no one would admit them into their families; and this aversion was carried so far, that they were excluded from the privileges of their fellow citizens, and even the doors of the temples shut against them. Under these circumstances, it appears to me very surprising that any man, or class of men, should voluntarily subject themselves to such obloquy on account of rearing hogs; yet these persons were so numerous as to be classed as a particular order in the state. But if, as is said, none of the eastern nations made use of the hog as an article of food, from what motive were such numerous herds of them reared both in India and Egypt? We often read in ancient history of one individual being in possession of a numerous herd. It may be said, that these hog herds consumed the animals themselves, but this is very improbable; and if they were deemed unclean and unwholesome food for the rest of the inhabitants of the country, it is not likely that such numbers of them could have been made use of. It is almost unnecessary to observe, that the swine was one of the animals forbidden to the Israelites by the law of Moses, and is accordingly held as unclean by the Jews to this day. Although permitted to Christians, yet there is still some degree of prejudice exists against them in the minds of many, from they know not what cause.

In addition to that very foolish one concerning the devil in the swine, they are said to be troublesome, and to have a spirit of contradiction about them, so that when you wish them to go one way, they show a disposition to go exactly the reverse. This is partly true; but, pray, would not a horse, were he not trained, prove equally obstinate and unmanageable? I have often observed the Highland ponies, when attempted to be laid hold of, after running at large upon the hills for a few years, as wild as any pigs. I have not the least doubt, that if the same perseverance were applied to tame and break in swine, as is used with other animals, there would not be so much reason for complaint. I have frequently seen a pig following a cottager from one neighbouring farm town to another, like a dog. That person probably had only one sow, and had been at some pains to feed it regularly, scratch, or stroke it down. With a little care, swine become so tractable, peaceable, and familiar, that I have seen them grunting around a fireside with the children, and seeming as happy with their companions as if rolling among the mire in a warm day.

On the smallest offence of this animal, the following is but too often the prevailing practice. The gudewife cries, 'Jock! look if ye see the swine.' He instantly obeys, and observes them to be among the potatoes, or in a field of growing corn. The hue and cry is raised; the gudewife bawls out, 'Bless me!

callans *rin!* for if ye'r father kens we'll a' be murdered. All the cotters' collies about the place run, as they understand what is to be acted. The dogs are all set upon a poor animal, some biting its heels, others hanging by its ears; hunting it out of one mischief into another, until the poor beast is all torn and besprinkled with blood! Its adversaries still pursue the chase, until the creature runs home for shelter; then the dogs go back for those which were left in the corn, and bring them home in a similar manner. The gudewife says to herself, 'The Diel got into the swine lang syne, and he'll bide in them. Rob! steik them up in the auld byre, and let them stan yoner, till they're dwaming for hunger, burst them! they're a daily cummer.' Next day they are turned out without a morsel in their belly, and run roaring mad from hunger, and will ere long be in another mischief; but who can blame them?

No person can expect to reap any advantage from such treatment of the animal. The fact is, they are accounted worth no care, and allowed to ramble up and down, without any proper person to look after them, excepting in the manner above mentioned. At any time, whenever a sow hears the name of a dog called upon, although she is lying quiet, she immediately shifts about for a more secure hiding place, with a terrible fright, and looks as wild as if it was going to be a high wind.

They are frequently half starved, which makes them the more ravenous. I see no animal about a farm that will be more peaceable and quiet than swine; as they are not inclined to ramble if they are well fed. Let any one try, and turn horses, cattle, or sheep, out without a keeper, and give them the same treatment as the swine, and then observe if they do not turn as unmanageable, do as much damage, and the character of the one become as bad as that of the other.

In some parts of Ireland, it is a common practice for a cottager to keep a sow, which both eats and sleeps in the house with the family. Another striking instance of the mild disposition of the animal came under my own observation lately. I purchased a young sow, of a good breed, nine months old, half gone with pigs. My youngest son, who is little more than three years of age, takes the advantage of her when lying asleep upon the dunghill, gets upon her, and gives her a slap upon the ear; in the act of rising he throws his leg over her, and will, in this manner, ride for half an hour or so about the farm yard, until she thinks proper to lay herself down. By this means she dismounts her rider, but at the same time seems as careful of hurting her jockey, as she would be about one of her pigs. He often allows her to share with him the bread or other meat that he has got in his hand.

I observed it mentioned some time ago, in the public papers, that a gentleman near London, trained his swine to run in his carriage, and drove four in hand through London.

Many other instances might be adduced, which I

have myself witnessed, but the above are sufficient. All the offences which swine commit are too often attributed to their bad disposition; but I am convinced, that it is entirely owing to bad management. Some may say, however, that many of them about a farm will be destructive, and I most readily agree that they will, if proper care is not taken to have secure conveniencies proportioned to the stock. Whatever number are kept, they do not much diminish any other of the farmer's stock, as they subsist mostly upon different food.

I shall conclude with observing, that it is rather strange, at this enlightened period, when agriculture, in all its branches, has arrived at such perfection, that the old custom of herding swine, both in this island and foreign countries, has been allowed to go so much out of use. It must be owned, that a swine stock, at present, are more worth attention than they were in those early ages, yet they are shamefully neglected, and allowed to scramble about for subsistence, picking a precarious and scanty supply from arable fields, under crop perhaps, or other sources, and in such a manner as to render their keep little profitable to their owner, and often a scourge to his neighbours.

PART II.

CHAP. I.

Recommendation to Gentlemen and Farmers to encourage the growth of HEMP *in this Kingdom, with some Hints on its Culture and Management.*

MANY able pens have been employed in describing the use and culture of hemp. You will find a very correct description of that plant in the ENCYCLOPÆDIA BRITANNICA, and in many other publications. The few hints now presented to the public, may be sufficient to encourage those to try the culture of this species, who may not have an opportunity of receiving information through another channel.

The cultivation of this valuable plant may be introduced into this country with great advantage, and the manufacturing of it would employ some hundreds, nay, I may say some thousands of the inhabitants, were it to become general. It is a mistake when it is said that hemp cannot be grown but upon a particular kind of soil. It is certainly very true that hemp, like all sorts of grain, has the most abundant crop on the best land. By a letter from a correspondent, from whom I have derived part of my information on this subject, I learn that he has grown some crops of hemp upon a light soil, and sold the produce so low at 6s. 6d. per stone, and that even then it paid better than corn. Hemp, however, at present brings a high price, and will likely continue to do so. If a Scots acre produce but only 30 stone, which quantity is little more than half a crop, it will still leave a handsome profit, after paying all expences, and the land will be left in a fine state for a succeeding crop of wheat. Hemp is by no means so impoverishing a crop as many suppose, and if the land is any way clean of roots, which it ought to be, there will be little danger of annual weeds growing amongst so close a crop. The land designed for hemp, which crop may follow corn if the land be clean, ought to be ploughed over after harvest, and about the month of February, or beginning of March, after being manured with good dung from the farm-yard, it may be ploughed in to a full depth. This manuring will be sufficient for a following wheat crop, and the hemp will be off the ground in time for the wheat being sown.

The hemp-seed, which should be of the preceding year's growth, is sown about the middle of May, if the weather is very warm, but even the beginning of June will answer. I should prefer sowing it when the weather is clear of frost, fine and warm. By that time the dung that was ploughed in will be pretty much rotten. The land should be ploughed into ridges not broader than nine feet, as one part of the hemp will be to pull out a considerable time before another is ready. When the ridges are not too broad, a person can go in each furrow and pull it, and yet not injure what is left.

The first season for pulling hemp is usually about the middle of August, when there is removed what is called the 'fimble hemp,' being that which is composed of the male plants. It would, however, be much better to defer pulling it for a fortnight longer, until those male plants have fully shed their farina, or dust, without which the seeds on the other plants will prove only empty husks. The male plants decay soon after they have shed their farina. One knows when they are ready for pulling by the fading of the flower, and when some of the stalks turn yellow. The second pulling is about Michaelmas, when the seeds are ripe. This is usually called 'carle hemp,' and consists of the female plants which were left.

Hemp must be pulled only when it is dry, and should be tied up in little sheaves, with a band near their top end, but not so as to touch the seed. The men that follow the pullers, after clearing the mould from the roots of the hemp, should set up the first sheaf with its roots spread so much as to enable it to stand upright of itself, and then set up four bundles, or sheaves, around it, with their heads leaning on the first, below its head, with their root ends standing out, or spread like a fan, so that they encircle it, in order to admit the free access of air. After standing a week or so, it ought to be turned about, so that every part may be rendered quite dry.

The best way to preserve the seed is to stack up the hemp, and thrash it out in the spring. It should be often turned, to prevent the mites, or insects, from damaging the seed. A few wood ashes dusted among the hemp would prevent them. After the seed has been thrashed out, the hemp must be watered, the same as flax. This operation is known to be finished by the reed separating from the bark. It must thereafter be spread, and lie such a time as may be deemed needful; then taken up, dried, and put into a house or stack, until it can be conveniently 'braked,' or broken.

The old way of separating the reed from the bark was by pulling out single stalks from the bundle and peeling them with the hand. A cloven brake will answer for breaking it. After this has been done, it can be switched, or swingled, with a wooden sword; but if it be a strong good crop, it will not need switching. The month of June is the best time to work the hemp, as the weather is then sunny and warm, which will make it work without drying by a fire. Those who have not the conveniences of mills, may peel, beetle, or break it, as they find proper.

About three Winchester bushels of hemp-seed will sow an acre of land. It should not be sown thicker. Last year I dunged about an acre of land, such as was prepared for potatoes after bearing two white crops, and sowed it with hemp-seed about the 1st of June. It turned out a better crop than I expected, from the dung not being put in earlier. I thrashed it off last spring, and sowed about the same quantity of land with part of the seed I preserved, and it has turned out an average crop. The hemp has been water-rotted, or steeped, &c. and put up dry and ready for breaking. It appears to be of a good quality. I intend this ensuing season to prepare five acres in the way described.

Let a farmer begin with one acre, and increase or diminish the quantity as he may find it answer. There is no fear of not finding a market for it, so long as there is such a demand for foreign hemp in this country. Whoever raises hemp or flax, should never attempt to work it the same season it is grown, as there are often accidents happening when drying it at a damp period with the fire. It is a better plan to stack it up when dry with the seed on, thrash it out in the spring following, water it in the usual way, then spread it, and when dry, take it up, and it will work with the heat of the sun in June. This is the safest and surest method, and the seed is thereby properly preserved. If the hemp-seed from your own crop should decrease in quality by sowing it too often, purchase foreign seed to supply its place.

The following rotation of crops, I apprehend, might answer a good purpose. First hemp, the land being prepared as described, next wheat, then beans, and dung again for a hemp crop, and so on. The land being dunged every third year will keep in good condition, and, of course, every one of these crops must be abundant; and as they are the most valuable and productive, it must undoubtedly make the farmer a better return than a rotation of turnips, barley, clover, and hay, then a white crop, or pasture. In this plan the land only receives dung every fourth year, and it may be kept equally clean in the one way as in the other. If it gets foul, the beans can be drilled, and receive both horse and hand hoeing, as well as the turnips. The hemp crop will in a measure destroy all the annual weeds.

There is one great advantage which wheat has after hemp—it is said to possess a property as a plant which renders it almost inestimable, viz. that of driving away all insects that feed upon other vegetables. In some places on the continent, the husbandmen are said to secure their crops from the mischievous attacks of such animals, by sowing a belt of hemp round their gardens, or on any particular spot which they wish to preserve.

Any farmer that has three small inclosures, or even one field divided into three equal portions, may carry on this hemp, wheat, and bean culture; and need not much encroach on his common mode of agriculture. If it is found to answer, a certain extent of land, according to the size of the farm, may be set apart for the purpose, so as not to interfere with the common practice, or diminish the cattle stock for want of fodder.

CHAP. II.

On the best method of Harvesting Corn in a wet and backward season.

In a wet and showery harvest, when there are a number of shearers employed, some of the corn will be liable to be cut wet, and often, when only a slight rain, the corn may, for an hour or so, not be in a state to bind up in sheaves. Frequently, from this cause, the reaping must be stopped for a time until the corn be dry. This breaks a labourer's day's work, or is time lost to the farmer. In a wet harvest, it would be wrong to make the sheaves so large as in good weather, and immediately after the reapers, they should be tied up loosely, and the band should be slipped up nearly to their top. They are to be set up singly, and have their butt ends spread out, so that they may stand upright of themselves. In this way the corn can stand a considerable time without receiving damage. It seldom rains so constantly but that there are some intervals of dry weather. After the corn has stood two or three days in the 'gait,' (it being so called in high districts where farmers follow this practice,) and when the gaits are thoroughly dry, which will not be very long in happening, as every sheaf has free access to the air, the bandsman must make one band when he begins, and apply it round the standing sheaf, turn it over, and tie it up. Then he should slip off the former band over the top, unloose it, and apply it to the next sheaf in the same manner, and so on through the whole of the field, every sheaf getting its neighbour's band. There is no other advantage from this way of using the bands for tying up the sheaves, than only from its being a little more expeditious. This mode is much preferable to a practice which prevails in many countries, viz. causing the reapers, when the corn is cut wet, spread it out with their hands very thin, and let it lie until they reckon it dry. More rain, or even the wetness which it had when laid down, makes it cling so close to the ground, that it will never dry in showery weather, nor yet well in dry weather without stirring it up. When the gaits are tied up as described, if the weather continue wet, set the sheaves up as usual in stooks, or in little round stacks in the field, of the shape of a sugar loaf, and put up by the hand, without any person going upon them to build them. This last method I would prefer. The stacklets may be made as high as a person can reach. The sheaves in them must hang in a sloping manner, with their tops up, so that the wet may run off.

In building, the upper sheaf must always be laid just to cover the band of the one below, which will naturally draw the stack, or 'tor,' (tower) to a point. It may be covered with a single bunch of straw, fixed on by two cross pieces of straw rope. If the corn is damp or wet, the tor must be so small, as not to require any heart sheaf; but if dry, it may be made larger; and perhaps even with a person on it in building it, and then it may have a few heart sheaves. In this way it will keep for months.

If the weather becomes any way good, it will dry it without any further trouble, as the stacks are made up so light, that a great part of every sheaf has the benefit of the air and sun. It may with safety be taken from these tors, and put up into the barn-yard in large stacks; but I still would recommend rather to keep them small in a wet harvest. In high lying districts, before the early oats were so common in the country, this was a general practice, the harvests being so late in the season. Oats or barley will answer very well to be put up in tors; but if wheat was to be gaited, I would recommend it to be put up in stooks after being tied up.

This method is only practised in a wet and backward harvest, or in high situations where there are frequent damps and fogs. In either case it will be found a great saving of labour, and preservation to the crop. Any person may make a trial of this plan, and if he finds it beneficial, he will, for his own interest, continue to adopt it.

CHAP. III.

On the great saving which would accrue to the Farmer by employing in part Oxen, in labouring the Farm, and especially in working the Thrashing Machine.

I hope the following observations on the advantage gained by employing *in part* oxen, in carrying on the necessary labours of the farm, will not be deemed useless: I mean, supposing a farm should require six horses to do the necessary labour, the gain from laying aside two of the horses and replacing them with two, three, or even four oxen. My reason for employing part horses and part oxen is, because the latter are not so well calculated for carting, especially when going journies to coal pits, lime kilns, &c. That the ox will draw as much as the horse cannot be doubted; yet, in this country, where the farmers use only single and double carts, the former animal cannot bear great weight on his back. In those countries, however, where they use waggons, the ox may be somewhat more useful for going journies, and perform draught better. He can only bear weight and take his draught from the top of his neck, which circumstance will account for the ancient custom of using the yoke, or a piece of timber laid on or over the necks of the oxen.

As to the saving, I shall first state the expence if

USING OXEN IN LABOURING A FARM. 97

the two horses are continued, which are proposed to be given up for oxen. Any person accustomed to feed horses, will allow that one shilling's worth of oats per day for a work horse, is no more than sufficient; and two horses to be fed thus for 40 weeks, is £.28. The small, or inferior oats, will in a great measure feed the remaining four horses, when they will not supply all the six. In the summer months, they may be fed on good pasture or cut grass, without being fed with oats. A horse now-a-days being kept at labour for a number of years, will decrease in value three pounds yearly, including accidents, which is, for two horses, six pounds. Add to this sum the £.28 for oats, and it will amount to £.34.—One need not enlarge on this head. It is well known to every farmer, that from the high prices draught horses are now purchased at, their value must decrease considerably on keeping them any length of time at labour. There are also many various disorders and accidents that horses are subject to, which every horse keeper and farmer must have more or less felt.

Let us now turn our attention to that other useful animal, the ox. Four oxen will cost rather less money than two good horses. The former will do the work of the latter with ease, because when two of them are employed in the plough, or carting manure, &c. for one half day, the other two will be idle and can take the place of the former two in the afternoon. The two that have nothing to do in the forenoon,

H

may work an hour or an hour and a half at the thrashing machine in the morning, and then go at liberty. They ought to be so employed alternate days.

It will be observed, that the four horses are sufficient to draw all the corn to market, and lime, &c. to the farm. On a possession of the size supposed, all the four oxen, when the horses are employed in distant carriage work, can be ploughing, taking out dung, leading in corn, hay, or turnips, &c. In such business oxen are as useful as horses. The former will plough in pairs, the ploughman driving or directing them with reins, the same as the latter, and are equally tractable. The advantages arising from keeping part oxen, are, 1. They are not liable to one disorder in ten that horses are. 2. Their increase of value; for they do not impair but improve, by working in the way described—which is little more than half employment. 3. Should an ox meet with any slight misfortune, so as to make him unable to do his daily task, there can be no loss, as he will answer the same for stall feeding as if sound; or were he to meet with any great or incurable accident, he will probably be sold for his first cost, as with so little work he must always be in pretty good condition. Should blindness or lameness befal any of his horses, where can the farmer pick up thirty, forty, or fifty pounds of outlay, in his purchase?

It is well known that straw, hay, and turnips, are all the food necessary for the ox when employed in the winter season, and in the summer good pasture or cut grass is sufficient. There is no occasion at any time to feed him with oats for his work; but if the farmer chuses to do so, it will make him in better condition for the butcher.

I stated the ox to improve three pounds yearly for three years; and when I did so, none can think that I exaggerated in my calculation. The farm oxen will thus advance twelve pounds annually, which, together with the greater expence in the keep, and decrease in the value of the horse, is a saving yearly of forty-six pounds sterling. This statement is only on a small scale, or refers to a small farm. Apply it to one on which ten horses are kept, which number may be reduced to six, and it will shew that there may be a saving of ninety-two pounds yearly, which is, in proportion to the extent of the farm, a still greater saving.

I know several gentlemen that have attempted to use oxen in the place of horses, but have condemned the plan and given it up. This want of success was owing to nothing but inexperienced or obstinate servants being ignorant of the mode of training them, or unwilling to take a little trouble to do that. I am fully convinced, that if oxen were properly trained, they would answer every purpose of husbandry; but if those who try the experiment do not set about it with a degree of prudence and perseverance, they will fail in the attempt, especially in a country where there is so much prejudice against the use of oxen. If they wish to ensure success, the masters must not only order a trial to be made, but they must attend and see the thing properly done, or perhaps do it themselves. There is not only a degree of unwillingness in some servants to try any new improvement, but there is frequently much stupidity and want of caution in the management of oxen when first put to work; and, what is worst of all, the impatience of the men sometimes leads them to abuse the animals on their showing at first some awkwardness, or even a disposition to be stubborn.

In order to make work more easy and familiar to oxen, I would recommend the same early pains to be taken with them as with horses. Those intended for work should be tied up when calves, or year-olds, a halter put upon them, trained to lead in hand, and when two years old, accustomed to be harnessed, and to draw an empty sledge, adding a little weight upon it by degrees. If the farmer has no old oxen to break them in with, let them be put into a plough for an hour or so at a time, with two tractable horses before them. Their draught must not be heavy at first; but when three years old, they will be able to perform every kind of work with ease, and go in a plough by themselves. Thus accustomed to labour by degrees, they will go through it more readily and steadily, and with less difficulty. In leading them about when young, and even when put to work, let them be kept always at a quick step, and they will be as speedy as horses, either in the cart, plough, or thrashing machine.

Some persons, perhaps, will not approve of this

plan of using oxen in carriage work; yet, I would advise such to try them in the thrashing machine, even supposing they put them to no other work. They are more steady in draught than horses, and the machine is very destructive to horses, especially in the spring season, when they must often thrash an hour or two in the morning, and immediately thereafter go to plough, hungry and weak from their not having time to feed. One set of oxen would answer, as in general they would not have above two hours of such work a-day. If the machine is already built, the step of the oxen may be too slow, if not trained to walk quickly. In that case, yoke them a little from the end of the lever, nearer the axletree than where the horses draw from; in this way, they would not have so much ground to travel over. The draught, no doubt, would then become a little heavier, but that circumstance would be a trifling matter to four or five oxen. Those who have machines to erect, should give them such a degree of velocity as would correspond with the step of their oxen.

I had a four-horse thrashing machine about fifteen years ago, but finding the inconveniencies the horses laboured under, I replaced them with four oxen. To answer the speed of the oxen, I worked them within the horse walk, so that a horse and an ox could draw a-breast. It was, of course, a heavier draught to the oxen than the horses; but the former had not so much ground to walk over, and went at a slow, steady pace. This answered the purpose to expectation.

About five years ago, I had occasion to erect another machine, the motion of which was calculated for two oxen. If the animals are kept from standing, and walk ever so slowly, the machine thrashes with great dispatch and effect.

I have wrought oxen occasionally for near thirty years, in the different employments here stated; but was frequently obliged to give them up, chiefly owing to inexperienced, cross, and ignorant men, who thought it below them to work with such animals. For my part, I should prefer oxen for moving a thrashing machine, to either horses, wind, or water. A water mill is often stopped from working by frosty weather, at a season when there is most occasion for its use; and a scanty stream is apt to dry up in a long continued drought in the summer, when the farmer should be thrashing out his wheat; whereas one pair of oxen are capable of thrashing 200 acres of crop with ease. The farmer need not trouble his head about collecting streams of water.

On a farm of this size four oxen may be kept with ease, which would thrive and pay the farmer equally the same as if they were going idle. Two hours work in a day would be only exercise for the animals, and make them eat their food with a better appetite. A two oxen power machine is quite sufficient to thrash 300 acres of crop. A large, heavy machine, is more expensive and difficult to work, and often out of trim. In this method there is no loss of time waiting for wind or water, or for the horses eating their feed betwixt the plough and machine, which would in the other be necessary. So much in favour of the oxen, which are at all times ready, besides the saving of expence, and their increase in value. But it is almost as possible to remove a mountain as to bring any new improvement into practice. Howsoever plausible and even reasonable such may be, its introduction must be a work of time. Many persons, even after they are convinced themselves that they would be benefited by an alteration, yet find it very difficult to remove the prejudice of others with whom they are connected, as has been already observed. Were it not for this consideration, I would have been disposed to recommend the further employ of oxen, which would make a proportionably still greater saving; *i. e.* to use one half oxen and the other half horses, in most farms where any number more than two teams are kept.

If any of my readers ever travelled through Glamorganshire, Monmouthshire, Somersetshire, Herefordshire, Devonshire, or Gloucestershire, &c. and have paid some attention to observe the working of oxen in these counties, such will not stand in need of much argument to satisfy them about the propriety of the plan proposed. The oxen in those counties are all taught to work a little when two years old. The method is, to put two oxen that have been accustomed to work *behind,* and two *before,* the young ones in the plough, for a few hours at a time. They are thus easily trained, and accustomed to keep step with the old ones. Not one of them out of ten ever gets sulky, unless they are not broken in until they are aged. When three years old, they are able to stand a full day's work along with the older bullocks, and generally work four or six in a plough, drawing a-breast in pairs. They generally have one pair of three year old ones annually taken into employment, and two of the oldest are turned out to be fattened. Working the young along with the old brings the former into their full step and tractableness, and appears to be an excellent way of training them.

At the present time, a pair of two year old bullocks, in the forementioned shires, will cost fifteen pounds sterling, and the six year old oxen are worth from thirty to thirty-five pounds per head. There too, the oxen work as long hours as horses, are as obedient to the word of command, and have as quick a step. Indeed the breed of cattle in that district are very proper for work, being tall and active. They are generally used in waggons, as well as in the plough; and, in short, they are able to perform the greatest share of the husbandry work in those and other counties in the west of England and south of Wales. The farmers there must find some advantage in keeping them in preference to horses, otherwise they would not do it. They have as much pride in their teams of oxen, as men in other counties have of their horses, and they take as great pains to have them well matched as gentlemen do about carriage horses. In working oxen they start early in the morning, and plough late in the evening, so as to let them have a long rest in the middle of the day. All animals can perform their labour best at a cool period of the day.

These and other considerations lead us to regret that the working of oxen has now gone so much out of use in the northern districts of Scotland. One should apprehend that the husbandman will again introduce them. The use of oxen in labour would be a double, nay, a treble saving at this time, of what it was formerly, owing to the enormous price of horses, and the expensive way in which they are kept. The oats they destroy are more than double their former cost. After working a few years, horses rapidly decay in value: On the other hand oxen will yield as much return for a year's keeping now as they were formerly worth altogether. Compare the profit on the one and loss on the other, and the advantages derived from the former must be very evident; yet the whole of the success will depend on the first training, or on accustoming them to a quick step and using them gently.

It would be superfluous to say much about the harness proper for oxen. Some persons still prefer the old custom of using the yoke, which was generally a piece of birch, or alder, about four inches square, and long enough to reach over the necks of the two oxen, with a staple and a ring in the centre of the timber to fasten that part of the harness which hooks in the plough. There are always two holes bored in the yoke for each ox, so as to receive an iron bow that goes up each side of the neck, and is fixed so as to connect and keep fast the whole. Taking out the bow sets the animals at liberty. Others prefer a collar nearly the same with that used for horses, only that it must be made very thick towards the upper part, and thin below; and the staple for the braces must be fixed higher than for horses, so that the top of the bullock's neck may catch the draught first, as it is the only place from which he can best exert his strength in drawing.

A FEW HINTS AND RECEIPTS,

FOR PREVENTING AND CURING DISEASES IN HORSES, CATTLE, AND SHEEP.

It has been already hinted, when speaking of the comparative advantages of horses and oxen, for the purpose of draught in the plough or wheel-carriages, that the former are liable to many severe disorders which greatly lessen their value to the husbandman. By having dry, clean, roomy, cool, or airy stables; by supplying them regularly with proper food and drink, or aliment suited to their age and work; and by making them undergo a due degree of exercise or labour (for too great exertions in work and perfect idleness, are nearly equally bad), much may be done to ward off diseases; and it is more the business of a farmer, as being both more easy and successful, to prevent, than to cure them.—It is not intended here to take particular notice of the various ailments to which the horse is liable, but only to offer a few observations on the remedies had recourse to for obviating or removing some of the more ordinary ones.

Of Sores and Bruises.

When horses, cattle, or any of our domestic animals are wounded, the treatment may be very simple, and much the same as in the human race. It is extremely improper to follow a practice not uncommon in many parts of the country among farriers, cow-doctors, and even shepherds—That of applying to the wound, or putting into the sore part, common salt, powder of blue vitriol, or tar, or cloths dipped in spirits, as brandy, rum, &c. or turpentine, or any other stimulant articles; for all such very much increase the pain, and by irritating the sore, may increase the inflammation even to the length of inducing mortification. Though the treatment may be varied according to circumstances, yet, in most cases, it may be sufficient to take notice of the following particulars. It will be proper to wash away any foulness or dirt about the part, and to examine particularly its condition. Should any large blood-vessel be cut and discharging copiously, it will be right to try to stop it by different applications, such as cold water, or touching it with spirits, to make the vessel contract; or putting over it some lint or sponge, with moderate compression, or bandaging at the same time, and not taking it off again for two or three days. Should the pressure fail of effect, caustic applications, such as the lunar caustic, or even the actual cautery, the point of a thick wire sufficiently heated, may be tried; or if a surgeon be at hand, the vessel may be taken up by the crooked needle full of waxed thread, or by the tenaculum, and then tied. Where there is no danger of excessive bleeding, and a mere division of the parts, a deep gash or cut through the skin or muscular substance, it will be right to adjust the parts, and try to keep them together by a strip of any common adhesive plaster; or when this will not do by itself, the lips of the wound, especially if it be a clean cut, may be closed by one or more stitches with a moderately coarse needle and thread, which in each stitch may be tied, and the ends left of a proper length, so that they can be afterwards removed when the parts seem to adhere. It is advised to tie the threads, because sometimes the wounded part swells so much that it is difficult to get them cut and drawn out, without giving pain and doing some mischief. If the wound be in a part that will allow a roller or bandage to be used, to assist in keeping the lips of it together, it may likewise be employed; for by supporting the sides of the wound, it would lessen any pain which the stitches occasion. With this treatment the wound heals often in a short time, or in a few days, rarely exceeding five or six, and sooner in the young and healthy than in the old and relaxed (providing there be no sickness, *i. e.* feverishness), and sooner in the quiet and motionless than in the restless and active. The use of a cool apartment, low feeding, bleeding and purging, which may be also proper, is to be afterwards considered.

Should, however, the wound be large, and inflammation, with the discharge of matter, likely to take place, it may still be proper to try, by gentle means, to bring the divided parts near to each other, and to retain them in their natural situation by means of a bandage. This should not be made too tight, but merely to the length of supporting the part. In this way, and by avoiding stimulant applications, the wound will heal more readily than otherwise, and the chance of any blemish following will be diminished. Washes of spirits, brandy, and the like, Friar's balsam, spirit of wine and camphor, turpentine, or any other such irritating applications, are highly improper, and sometimes make a fresh clean wound, that would readily heal almost of itself, inflame, and perhaps mortify or become a bad sore. In old, foul, sluggish ulcers, no doubt, such applications may, at times, be found requisite, especially in animals of particular constitutions, and where the sore has arisen spontaneously or has not been produced by any wound.

Over the whole sore, or where the part is bruised, as well as cut, or where there is a tendency to suppuration, a poultice should be applied and kept on by suitable bandages. The poultice may be made of any kind of meal, fine bran, bruised lintseed, or of mashed turnips, carrots, &c. The following has been found useful as a common poultice: 'Fine bran one quart; 'pour on it a sufficient quantity of boiling water to 'make a thin paste; to this add of lintseed powder 'enough to give it a proper consistence.' Such poultices may be kept applied both to wounds that have had their edges brought together in any of the ways mentioned, or to large ones which cannot be so treated, but must be left open till they heal by the formation of new flesh and skin. In very extensive wounds, or where the part is torn open and a loose flap of skin left, after carefully examining the part, washing it with tepid water, and removing any dirt or foreign substance, it will seldom if ever be proper to cut off any of the skin, but it should be accurately replaced, and then there should be applied over it a large soft poultice, milk warm. At first nothing more than this should be done to the wound; for no stitches would keep together a large one, especially if torn, and as the parts swell, the points fixed by the stitches may be over-stretched, inflame, and mortify. The poultice may be kept on for a week or ten days, or even longer, if necessary, changing it once or twice a-day, and cleaning the wound, when the poultice is removed, by washing it by means of a soft rag or linen cloth (some sponges are too rough for this purpose,) with water not more than blood warm; or, where the wound is deep, the water may be injected into it by a syringe, in order to clean it from the bottom.

In the case of severe wounds, attention should be paid to the condition of the animal in other respects. There being always when such happen a tendency to violent inflammation and fever, that may end fatally, means should be employed to moderate both. The apartment should be cool and airy, and so quiet that the animal should not be disturbed; the drink should not be warm but cold rather, and given freely, though not in too large quantities at a time; the food should be sparingly given and of a poorer quality than usual, and should be rather succulent and laxative, than dry or apt to produce costiveness; bleeding may be employed either generally from a vein, or, in some cases, when it can be done, by cupping from the hurt part in the case of a bruise (though this last will seldom be requisite or found convenient), and it may be done more than once or twice, as may seem proper; laxative medicines also ought to be given and repeated, as there may be occasion. What relates to such will be afterwards considered.

In the course of a few days, when the wound, by care and proper management with the poultices, begins to put on a healthy appearance, and seems to be clean and of a reddish colour, not black or bloody, then there may be applied an ointment made of tallow, lintseed oil, bees wax, and hog's lard, in such proportion as to make it of a consistence somewhat firmer than good butter. The ointment should be spread on some soft clean tow, and when applied to the wound, it ought never to be tied hard upon it, which is done too frequently and very improperly, as over-tightness retards the cure; but only fixed by a bandage of a proper length and breadth, (for a mere cord is often improper,) so close and securely as to keep it from slipping off. This application may be changed once a-day, or when nearly well, and discharging but little, once in two days.

In all cases there will be required a shorter or longer

time to effect a cure, after this safe and simple way; but when the wounded part begins to discharge a whitish, thick matter, and is observed to fill up, the general treatment and dressings to the sore, now mentioned, should be continued; and in the course of the cure, the animal, when free of fever, may be allowed better provision, and may get some exercise of a sort, and to the degree, suited to it. Nothing more, in most instances, will be necessary; for in a healthy body, any accidental wound, unless large, soon heals. If, however, the animal be feeble, from the loss of blood originally, or from the long continuance of a feverish state produced by the inflammation attending the wound, or from weakness arising from confinement, or connected with its constitution naturally; and if the wound appear to be in a stationary state, very pale and flabby on its edges, with a thin discharge, then in such a case, better food may be given to it, and if still no change should be observed, along with the better food, the wound may be treated somewhat differently from what has been already advised. The ointment may be made more stimulant, by adding to it some resin and less bees wax, or what would be more stimulant still, some common turpentine; for it is only in very rare cases that oil of turpentine can be requisite. The effects of an alteration in the mode of treatment should be particularly marked, and stimulants should be laid aside, continued or increased, according as may be judged proper. Perhaps, before changing the dressings applied to the wound, or before rendering them more stimulant and active, by using heating applications, the effect of closer bandaging may be tried; for sometimes by keeping the parts a little more firmly together, the cure is promoted.

In the case of very large wounds, that have ragged edges, or where the part has been torn as well as cut, it may be proper (after cleaning away any dirt or foreign matter by washing it well with water blood warm, and after getting the bleeding stopped, if necessary,) to bring the sides, or lips, as close together as can well be done by the use of gentle force, and to preserve them in a proper situation by means of a suitable bandage. The sore should afterwards be examined once a-day, and any dry, hard blood, or matter, should be cautiously softened and washed off by warm water. If the wound is very deep, and discharges any black, corrupted matter, the water may be thrown into it by a syringe, at the time it is dressed. At this period, that is when the tow and cloths are to be renewed, in the case of such wounds fomentations may be of use. Indeed, wherever there is more of a tear or bruise than a clean cut, or where there is even only a severe bruise, fomentations may be used either more constantly or occasionally, according to circumstances. Unless in the case of internal pain as for relieving gripes, fomentations (which consist in applying woollen cloths dipped in hot water and then hard wrung out, or, instead of hot water, in hot decoctions of bitter or aromatic herbs,) should never be made warmer than blood heat, as otherwise they may rather inflame a sore or bruised part. They are sometimes made with expensive or difficultly found materials, and very commonly by boiling mugwort, wormwood, mint, southernwood, tansy, camomile flowers, &c. and then applying cloths dipped in the decoction and quite wet, or without being properly wrung out, as hot as the animal can bear them; but plain water will answer the purpose perfectly well. Fomentations are found to be most effectual when frequently repeated and used for some length of time, and rather under-warm, in the case of any external injury.

If the animal be in full health, and have not suffered the loss of much blood from the wound, he would require to be bled in most cases, in order to diminish the fever arising from the inflammation attending the wound. A purging draught, or ball, is advised by Mr White of Exeter, to be given as early as possible, and the horse's diet to be confined to succulent food, such as bran mashes, along with hay. He may then be allowed to drink pretty freely and often, but he must be kept perfectly at rest*.

---

* This intelligent writer, late Veterinary Surgeon to the 1st or Royal Regiment of Dragoons, remarks, that, 'When this plan is adopted, the inflammation, swelling, and fever, which always follow an extensive, lacerated wound, will be much more moderate than it would otherwise have been, and in a few days will have subsided considerably; a white matter will then flow from the wound, and the horse will not appear to suffer much pain. When this has been accomplished, it is necessary to endeavour as much as possible to bring the divided parts together, and

On the use of Laxatives and Purgatives for Horses.

These form an extensive class of medicines, and are given to horses for various purposes. The for-

---

there will then be less danger and pain from drawing the bandage with more force for this purpose.' Great tightness, however, should be avoided, as improper. 'Warm water,' Mr White says, 'may still be used for cleansing the wound; but when the inflammation is quite gone off, some stimulating liquids may be employed, but these are unnecessary when the divided parts can be brought into contact. When this cannot be effected, or when there is a loss of substance, the wound cannot heal without the formation of new parts, and stimulants are often required to accelerate this process. At first, the weaker preparations are to be used, such as dilute spirit, (whisky or brandy and water,) or a weak solution of blue vitriol: but when the healing process goes on slowly, the matter becoming thin and losing its white colour, the stronger stimulants, as tincture of benzoin, or even oil of turpentine, may be applied, and the constitution invigorated by a nutritious diet, such as malt and oats, or carrots: And when the discharge is very considerable, and appears to weaken the animal, that is more particularly necessary, and must be assisted by medicines of the tonic kind, such as Peruvian bark, cascarilla, vitriolated iron, and sometimes porter or beer, and even opium. It is only in very deep and extensive wounds, however, when there is a profuse discharge, and constitutional weakness, that this treatment is required.

'When wounds of this kind terminate fatally, it is generally from the violence of the inflammation and symptomer act in a gentler manner than the latter, and are given occasionally to remove costiveness, or for the

---

matic fever causing gangrene, delirium, and total exhaustion. Our first and principal object, therefore, should be to restrain this inordinate inflammation, by every means in our power; but farriers, ever in opposition to nature, generally destroy their patients in these cases; torturing the unfortunate animal by the application of violent stimulants, and even caustics; cramming into the wound hard tents, and persuading their employers that this cruel and absurd treatment will infallibly heal the wound. When we have succeeded in these extensive lacerated wounds, so far as to bring on a healthy suppuration, a discharge of white matter, and an appearance of new flesh sprouting up in various parts, in small granulations of a red colour, we may be satisfied that the danger is over.

At this period we may safely use more force in bringing the divided parts together; and if the wound appears languid, wanting that red appearance we have just described, and discharging thin matter, some of the stimulants we have mentioned may be employed: Still it is improper to cram tents into the wound, or daub them over with stinking ointments. If the red granulations form so luxuriantly as to rise above the level of the skin, they must be kept down by red precipitate, burnt alum, or other applications of this kind; pressure will also be effectual on this occasion, laying a piece of soft lint on the part, and confining it with a roller. Should the sides or edges of the wound become callous, caustics must be applied to remove the old surface, and then fresh attempts should be made to bring them into contact.'—*Treatise on Veterinary Medicine*, Vol. I. p. 310, &c.

purpose of aiding the expulsion of worms, or on changing the food of the animal, that is, to prepare him for entering on the grass, or for getting wholly succulent food, or more frequently, in other cases, when he has been taken into the house and put on hard food, and so may have become over-costive in his belly, and thence apt to be less thriving, which condition is indicated by a dry, staring, or rough coat, or cover of hair. Laxatives are meant merely to operate on the intestines in a gentle manner, and to occasion the evacuation of their contents without any prolonged or violent discharge. Some persons use castor oil; but this is somewhat expensive, and may not in all cases be at hand, and olive oil, or lintseed oil, has nearly the same effect. The dose of castor oil may be from half a pint to near a pint, and that of the others about a pint and a half. Some persons recommend the following mixture, viz. 'common oil three 'or four ounces; glauber salts, or sulphate of soda, 'four ounces; and about the same quantity, or a little 'less, of common salt; the salts being dissolved in wa- 'ter, or rather water-gruel, blood warm, in the propor- 'tion of a quart, or chopin.' Two thirds of this mixture are given at first, and after waiting an hour or two the remainder.

Mr White says, 'When a laxative ball is required, 'the following will be found useful: Aloes half an 'ounce, Castile soap three drams, and syrup enough 'to form a ball, for one dose.'

Purgatives are more active than laxatives. They

are given in feverish or inflammatory disorders, to relax the vigour of the constitution, diminish the frequency of the pulse, or force with which the blood circulates, and so to lessen the fever. They are likewise used, either more constantly or only occasionally, to expel worms and bots, of which last there are two or three distinct sorts that infest this useful animal, but which it would be needless to particularize here. They are described by Mr Bracy Clark, in the Transactions of the Linnæan Society at London, and seem to be the larvæ of the *oestri equi*, or the *oestri hæmorrhoidales*, both of which, though in different parts, are apt to lodge in the stomach and intestines, especially the last, or great gut of horses. In a laxative state of the bowels, any how induced, many of these vermin are discharged. How this happens, whether they are expelled mechanically by the forcible contraction of the gut, or whether they are otherwise injured by the acid nature of the laxative, it may be difficult to explain. Medicines exhibited under the notion of poisoning them, have generally very small effect.

Mr White of Exeter, in his Treatise already referred to, has also delivered some good remarks on worms, at considerable length, to which I beg leave to refer.

Sometimes not only laxatives, but active purging medicines, are used for the purpose of removing colics supposed to arise from hardened fæculent matter in the bowels, which remaining there occasions obstruction and pain, or gripes. All animals, indeed, subject to costiveness, are apt, more than others, to have such collections formed in time, in the course of the intestines; and of all our domestic animals, horses would appear to be most liable to this description of disorder. The horse has a good length of intestines, which have many pouches, or sacs, and fæculent, hard balls, or 'chalky concretions,' as they are by some called, and rather erroneously thought to be, may be formed in different parts, but they are generally in the lower portion of what is similar to the colon in the human race. Accordingly, it is mentioned in page 57 of the late Mr James Clark's publication 'On the Means of Preventing the Diseases of Horses,' that such concretions are apt to be produced in the intestines of horses which have been long used to dry hard feeding. In that page Mr Clark speaks of the use of early, or spring grass, as a laxative to cleanse the bowels, and carry off the concretions. It is probable, that they frequently take their origin in different sacs, or wide places, in their intestinal canal, in which they roll about and accumulate by the accretion of new matter, till, from the motion of the bowels, they start out into the common or straiter passage, and there, by reason of their bulk, give rise to disorders more or less severe according to circumstances. Either they may produce temporary gripes, which may depart on their rolling into some new or open pouch, or they may remain fast, close up the passage, occasion inflammation, and kill the animal*.

---

* The balls are often surrounded by a mass of fæculent

Probably, indeed, more horses die from this or other effects of costiveness, than we have yet any notion of. There are many old worn out racers, hunters, and carriage horses, that are destroyed by its effects, and many more of them would die by it than do, did not the more succulent coarse food, which they generally get when they fall into the hands of those who afterwards purchase them, contribute somewhat to preserve them*. When obstructions are supposed

---

matter, but on being examined, a nucleus, or a first hard ball, will be found in the middle of them. Indeed this middle portion, formed generally of concentric layers, often becomes so hard as to resemble a stone, some kind of agates, and what is curious, it would appear from some late accounts, that they contain a good deal of animal matter, with comparatively little of any vegetable substance mixed with it. They would appear to be formed somewhat after the manner of gall stones.

* Mr Laurence, in his 'Treatise on Horses, and on the Moral Duties of Man towards the Brute Creation,' Vol. II. p. 470, when treating of colic, a disorder fatal to many horses, one, indeed, which according to some, 'kills more horses than all others put together;' says, 'The primary cause of a common fit of the gripes in a horse, is, nine times out of ten, an accumulation of indurated excrement in the intestines; for, independently of the solid obstruction so occasioned, the usual proximate causes would seldom have power to work those serious effects we witness: Thus, in a horse, the colon of which was not previously infarcted and plugged up, the effect of a slight cold thrown upon the bowels, or the devouring a few new beans, would probably pass off with a very moderate struggle from nature.'

to arise from indurated fæces in the intestines, and a high degree of disorder is produced, and the horse unable to get any passage in its belly, after repeated attempts to do so, it may be proper not to try to force a passage by giving purgatives, until it has been ascertained whether the hand could reach their situation, and remove them, a thing which is seldom the case until the effect of a charge or injection has been tried, and a large common clyster of water, blood warm, in the quantity of from 3 to 4 quarts, with a small handful, or about five or six ounces of common salt dissolved in it, will serve the purpose as well as any other.

As all material changes in the condition of animal bodies, should take place in such a way as not to shock the constitution, or general health, by their greatness, or suddenness; so it is proper, which, indeed, is attended to in most cases by intelligent grooms, to put the horse under some preparation for physic. This is done generally, in such a way as to prepare the bowels for receiving the purgative, by supplying him with some succulent food, or mashes of bran, &c. for a day or two before it is given. This favours the operation of the medicine, and prevents so high a degree of irritation being produced by it as might happen were the intestines less full, and their contents in a dry state. A peck of bran, given in three or four mashes, in the course of twenty-four hours, will answer the purpose; and his drink may be in larger quantity, and more frequently given, than usual, with a less proportion of hay. Some add to the bran, in order to keep up the strength of the animal, a pound or two of bruised oats or oatmeal.

The morning is the usual time for giving the medicine, or after the horse has fasted for two or three hours. A small quantity of water slightly warm, may be given the animal after it has received the medicine, which will promote its operation. In the course of the day, he should receive, in the stable, mashes of bran, and a full allowance of water, given at intervals, not, however, quite cold, and the proportion of hay should be moderate. Next morning the horse, properly covered with clothing, should be taken out to receive some exercise, when the medicine will operate more or less freely. If a walk does not encourage its operation, a gentle trot may be tried. If purging have ensued to a full degree when first taken out, he may be allowed to remain in the house during the rest of that day. The day thereafter, when the purging is generally over, the horse may be allowed his usual food. If he is to get hay and corn, and is to be prepared for hard exercise, the return to the dry food should be gradual, or a full portion of bran may be substituted for a part of the usual allowance of corn, and a more liberal supply of drink during the first three or four days after the physic.

As to the kind of medicine proper to be given as a purgative for horses, though there are many forms or prescriptions in use, yet it would appear to be confirmed in the experience of the most accurate and skilful practitioners, that those in which aloes make the chief ingredient, are the best, the most safe, and certain in their operation. Of the two kinds of aloes in ordinary use, the Barbadoes and Succotrine, the former has of late years rather obtained preference, as being, though stronger, less variable than the latter in its effects. The aloes is generally mixed or beat up with Castile soap, and such a quantity of syrup made of sugar and water, as will make a moderately firm bolus, or rather ball, of a proper consistence. For the most part, some other articles are likewise found in the composition purging balls, such as aromatic substances, like powdered ginger, or essential oils, as of carroway or anise seeds; and such are reckoned of use to make the physic sit lighter on the stomach, or to diminish the risk of its producing sickness or gripes. If such are not made of too expensive cordials, they may be conjoined with the other, for unless added in too large a proportion, they can do no harm. The proportion of aloes used must vary according to circumstances, the kind, and age of the animal, &c. Care should be taken not to make the physic too strong, as irritation, or even violent inflammation of the intestines, has been the consequence of doing so in many cases. This is particularly necessary when a purgative is first given to any animal, or even where a new kind of purgative is given to an animal accustomed to get physic. If the bowels be weak and irritable, and apt to be lax, the dose should be gentle. The proportion of aloes in a ball may be from half an ounce to an ounce; and of Castile soap, about two-thirds of the quantity or weight of the aloes, with a dram of powdered ginger, and as much syrup as will be requisite to make it up*. When they are prepared, they

---

* Mr White, already mentioned, whose experience as a Veterinary Surgeon was extensive, and whose observations on the diseases and treatment of horses, are valuable, gives the following receipts for purgative balls. No. I. 'Barbadoes aloes, 5 drams; prepared natron, 2 drams; aromatic powder, 1 dram; oil of carroways, 10 drops; syrup enough to form a ball, for one dose.'

No. 2. 'Barbadoes aloes, 7 drams; Castile soap, half an ounce; powdered ginger, 1 dram; oil of carroways, 10 drops; syrup enough to form a ball, for one dose.'

No. 3. 'Barbadoes aloes, 1 ounce; prepared natron, 2 drams; aromatic powder, 1 dram; oil of anise-seed, 10 drops; syrup enough to form a ball, for one dose.'

'The ball, No. 2. I have generally found sufficient for strong horses, and have scarcely ever had occasion to go farther than No. 3. Should any one, however, be desirous of a stronger medicine, it may readily be procured by adding one or two drams of aloes, or one dram of calomel, to the ball No. 3.: but I must not omit to observe, that there appears to me to be a considerable danger in making the addition.' Vol. I. p. 229. He says also, ' I have for several years employed the Barbadoes aloes very extensively, giving often from thirty to fifty doses in the course of a week; and have found that from half an ounce to one ounce may be considered as the proper dose. For a delicate blood horse, half an ounce generally proves sufficient; to a common saddle horse, five or six drams; to a waggon horse, 1 ounce.' p. 227.

should not be kept so long as to become over-dry; for then their operation is apt to be uncertain and irregular. Balls are often given in an improper, slovenly, and painful manner. Some grooms are dextrous at the business, and after seeing it once done by an expert person, nothing more is wanted than a little attention and address, with a few trials, to do it in a proper way.

On the Cure of Gripes in Horses.

This disorder goes by different names, in different districts of the country, as *fret*, from the uneasiness attending it; *bots*, from its being thought to arise from these animals or worms, &c. The symptoms of gripes, or pain in the bowels of horses, occurring in flatulent colic, are generally known, and need hardly be mentioned here. The animal looks dull, and rejects his food; becomes restless and uneasy, frequently pawing; voids his excrements in small quantities, and often tries to stale; looks round, as if towards his own flank, or the seat of complaint; soon appears to get worse, often lying down, and sometimes suddenly rising up, or at times trying to roll, even in the stable, &c. As the disorder goes on, the pain becomes more violent, he appears more restless still, kicks at his belly, groans, rolls often, or tumbles about, with other marks of great agitation, becomes feverish, and has a cold moisture at the roots of the ears, and about his flanks, and when he lies at rest a little space, begins to perspire strongly, and to get covered with sweat, more or less profuse.

The disorder may be induced in different cases from a variety of causes, such as indelicate or washy horses; from too sudden exposure to cold, as on taking them out of a hot stable and letting them stand for some time in the open air without exercise; or from drinking cold water, especially if in large quantity, and when warm from labour or otherwise; or from taking food that is less digestible than what the animal has been accustomed to, as bad hay, or even bad corn; over succulent, green, laxative food, especially on first using it; or sometimes from an accumulation and confinement of air, produced from the fermentation of the food, or, perhaps, in a way not well ascertained; which air is apt to distend the stomach and intestines, giving rise to pain, spasmodic affections, and sometimes to inflammation of them: The quantity of air produced in some rare cases, has even occasioned a rupture of them. In most cases of ordinary gripes, signs of flatulence, or of the presence of air confined in the bowels, occur and constitute a part of the disease, or increase it. The removal of it is, therefore, an object to which the attention of most grooms has been in a chief degree directed; and as it can frequently be gotten rid of, and the disease cured, by exciting the powerful action of the intestines, cordial and stimulating medicines are had recourse to, and, no doubt, in many have afforded relief. Some farriers, indeed, without much care in distinguishing cases, almost exclusively rely upon such, and employ them too freely. This, however, should not be done; for it sometimes happens, that disorders not unlike flatulent colic or gripes do occur, when there is neither pent up air present, nor any relaxation, or want of energy and action in the intestines themselves, and stimulating medicines might then do no good, but often much mischief.

It should be observed here, that in some instances the disorder departs after a longer or shorter time, even when no remedies at all have been employed; but in others, the use of such has shortened or alleviated it, or has prevented the fatal consequences which at times attend it.

When the disorder is early discovered, or has newly come on, it will be proper to lose no time to get ready a clyster, and likewise a medicinal draught for removing the wind, and abating the pain. After removing with the hand any excrement in the last or great gut, that can be reached by it, a clyster made of five or six quarts of water, or water-gruel, blood warm, and six or eight ounces of common salt, may be injected; and one or other of the following draughts may be given, before or about the same time; viz.

No. 1. Take of Venice turpentine one ounce, beat it up with the yolk of an egg, and then add of peppermint water, or even of common water, if the other is not at hand, one pint and a half (English measure), and two ounces of whisky or gin. This will serve for one dose.

No. 2. Take of table beer, a little warmed, one and a half pint (English); common pepper or powdered ginger, one tea-spoonful; gin, whisky, or rum, from two to four ounces, or from one to two glasses full; these mixed together for one dose.

No. 3. Oil of turpentine one ounce, and water-gruel one and a half pint (English) mixed, for a dose.

These preparations are simple and efficacious. The last is much the same with a mixture recommended by Mr White in the following words. '*Antispas-*'*modic Mixture for flatulent colic.* Oil of turpen-'tine, two ounces; cold water-gruel, one pint: Mix 'for one dose. To an inexperienced person, this 'might appear a very formidable remedy; but it is 'not only very safe, but seldom fails of giving relief: 'Many practitioners give it in larger doses.' *Treatise on Veterinary Medicine*, Vol. II. p. 191. These and the like preparations, may be given either out of a bottle, or drench-horn, one or two persons raising and keeping properly up the horses head, while another who administers the medicine, pulls out, and a little aside, the tongue, with his left hand, and with the other pours in the draught.

Cordial drenches of the kinds recommended, No. 1, 2, or 3, with the clyster, will have effect in ordinary cases to relieve the disorder. But should this not be the case, after waiting an hour or two (longer or shorter according to the severity of the ailment, or the period since its commencement,) then the medicine should be repeated, but in a less dose than at first— perhaps one half or two-thirds of the former quantity. The horse should be occasionally walked out, properly covered with clothes, lest the chill air bring on shivering, and give rise to feverishness; and his belly should be now and then rubbed a considerable time at once, five or ten minutes, but with intervals of rest, so that it may have time to stale or dung. If the disorder does not yield to these remedies, then others must be employed of a more active nature. Some persons recommend castor oil in the proportion of half a pint to a pint (English,) with an ounce or two of laudanum, or tincture of opium, mixed with water-gruel, in the quantity of a pint English or rather less. In case the horse has lain down, and continued so for some time, and is covered with sweat, when he rises two or more persons should be employed to rub him dry, and he snould also be kept well clothed. The stable should be airy, moderately cool, and his place in it roomy and well littered, to keep him from hurting himself should he roll about.

Draughts of liquid medicine operate more speedily than any other form; but as the disorder may attack a horse during a journey, where such cannot readily be procured, Mr White has given a receipt for a ball for the convenience of those who travel; and if it be wrapped up closely in a piece of bladder, it may be kept a considerable time without losing its power. The ball is composed of the following ingredients, viz.

'Castile soap three drams, camphor two drams, ginger one dram and a half, and Venice turpentine six drams: To be made into a ball for one dose.'

Laudanum may be used in cases of urgency, especially in the wet or lax gripes. Take a quart of beer, and make it a very little warmer than blood heat; then put a table spoonful of powdered ginger into it, and a small wine glassful of laudanum, just before it is given to the horse. This, in most cases, will give ease in a short time; but if the complaint is exceedingly violent, give about half the above quantity again in fifteen or twenty minutes. As soon as the pain seems to be abated, if the belly is costive, give the horse a purgative. In case of looseness no purgative must be given, and the laudanum, which is of a binding nature, will correct it.

Among the medicines which are employed to afford relief to horses in gripes, or to animals of any species when suffering pain, opium or laudanum (which is a solution of it, containing in 25 drops of laudanum about one grain of opium) is the most powerful. It must, however, be given with due caution, as pain may arise from causes, the effects of which the use of opium would rather tend to aggravate than assuage. Accordingly, it is well observed by Mr White, that 'When *pain* is occasioned by *inflammation*, it is seldom proper to employ opium, or any medicine of that kind; but when it depends upon *spasm*, or *irritation*, no medicines are so beneficial. In inflammation of the bowels, for example, opium would certainly do much injury, but in flatulent or spasmodic colic, or gripes, it seldom fails of success.'—*Treatise on Veterinary Medicine*, Vol. II. page 187.

This intelligent writer, mentions for horses affected with such a colic, or where the use of anodynes are requisite, the following preparation: Namely, ' opium ' one dram, or sixty grains; Castile soap, two drams, ' and powdered anise-seed, half an ounce or four ' drams; to be made into a ball with syrup, for one dose.' Perhaps here the quantity of opium is fully large, and if there were only three fourths of a dram, the proportion would be better adapted to the size of most horses, and even half a dram may be sufficient, if it be meant to repeat the dose in the course of twenty or thirty minutes *.

---

* In speaking of the medicines for gripes, or the flatulent colic, sometimes termed *fret*, Mr White mentions ' A few domestic remedies, which may be employed when medicines cannot be procured in time. 1*st*, A pint of strong peppermint water, with about four ounces of gin, and any kind of spice. 2*d*, A pint of port wine, with spice or ginger. 3*d*, Half a pint of gin diluted with four ounces of water and a little ginger. I have seen the complaint removed by warm beer and ginger, or a cornial ball, mixed with warm beer.' He repeats the caution given respecting the necessity of distinguishing the flatulent, or windy, or spasmodic colic, from the inflammatory one, and from that which depends on costiveness, and observes, that, ' It is *always* necessary to empty the bowels by means of glysters; and, should the horse have appeared dull and heavy,

Receipt for Curing a Surfeit or Cough Bad Coat in Horses.

Take crocus metallorum, or liver of antimony, one ounce; sprinkle it to make it stick, or mix it with moist bran. This may be given to horses subject to this disorder once a-day, among their oats; it relieves the appetite, destroys worms, sweetens the blood against all obstructions, opens the passages, and improves tired and lean horses in a great degree; it is also of great service in coughs and shortness of breath. It may be given daily from two to four weeks, and will soon produce a fine coat. The horse may be worked while he is taking the medicine, care being taken not to expose him to wet or cold.

Receipt for the Garget, or Udder Clap, to which Cows are subject after being put from giving Milk.

This disorder is very frequent in cows; it affects the glands of the udder with hard swellings, and often previous to the attack, it will be advisable to bleed. If costiveness attends it, give a laxative drench after the paroxysm, which will prevent its return.'—*Treatise on Veterinary Medicine,* Vol. II. p. 192, 208. Vol. I. p. 92.

arises from the animal not being clean milked. It may be removed by anointing the part three times a day with a little ointment composed of camphor and blue ointment. Half a dram or more of calomel may be given in warm beer, from a horn or bottle, for three or four mornings, if the disorder is violent.

Receipt for making a Cow take the Bull.

Mix a small handful of spurge in some hay and give it to the cow to eat, and it will produce the desired effect in a few days.

For Curing the Redwater in Cattle.

Take an ounce of bole Armeniac, half an ounce of dragon's blood, two ounces of Castile soap, and one dram of Roche alum. Dissolve these in a quart of hot ale, or beer, and let it stand until it is blood warm; give this as one dose, and if it should not have the desired effect, give the same quantity in about twelve hours after. This is an excellent medicine for changing the water, and acts as a purgative when they are dry in the dung, which is often the case when the water is red. I have practised this method of cure for forty years, with great success. Every farmer that keeps any number of cattle, should always have some doses of it by him.

For the Foot Rot in Sheep.

FIRST CURE.

Take a piece of alum, a piece of green vitriol, and some white mercury—the alum must be in the largest proportion; dissolve them in water, and after the hoof is pared, anoint it with a feather, and bind on a rag over all the foot.

SECOND CURE.

Pound some green vitriol fine, and apply a little of it to the part of the foot affected, binding a rag over the foot as above. Let the sheep be kept in the house a few hours after this is done, and then turn them out to a dry pasture. This is the most common way of curing the foot-rot in Middlesex.

THIRD CURE.

Others anoint the part with a feather dipt in aquafortis, or weak nitrous acid, which dries it at once. Many drovers that take sheep to Smithfield, carry a little bottle of this about with them, which, by applying to the foot with a feather, helps a lame sheep by hardening its hoof, and enabling it to travel better. Some may think aquafortis is of too hot a nature, but such a desperate disorder requires an active cure, which, no doubt, is ever to be used cautiously.

FOURTH CURE.

I have sometimes used the following means with success. Spread some slacked quick lime over a house floor pretty thick, pare the sheep's feet well, and turn them into this house, where they may remain for a few hours, after which turn them into a dry pasture. This treatment may be repeated two or three times, always observing to keep the house clean, and adding a little more quick lime before putting them in.

I have frequently heard the Yorkshire farmers observe, that the only cure for this disorder in sheep was to cut off their heads at once. The disease, no doubt, is very destructive, but some of the foregoing medicines, with proper attention, will effect a cure. The foot must be often dressed, and the sheep kept as much as possible upon dry land. Those animals that are diseased should be kept separate from the flock, as the disorder is very infectious.

How to Prevent Sheep from Catching Cold after having been Shorn.

Sheep are sometimes exposed to cold winds and rains immediately after shearing, which exposure frequently hurts them. Those farmers who have access to the sea, should plunge them into the salt water. Those who have not that opportunity, and whose flocks are not very large, may mix salt with water and rub them all over, which will in a great measure prevent any mishap befalling the animal after having been stript of its coat.

It is very common in the months of June and July, for some kinds of sheep, especially the fine Leicester breed, which are commonly thin skinned about the head, to be struck with a kind of fly, and by scratching the place with their feet, they make it sore and raw. To prevent this, take tar, train oil, and salt, boil them together, and when cold, put a little of it, on the part affected. This application keeps off the flies, and likewise heals the sore. The salt should be in very small quantity, or powdered sulphur may be used instead of it.

For Curing and Preventing the Scab in Sheep.

Take one pound of quicksilver, half a pound of Venice turpentine, two pounds of hogslard, and half a pound of oil, or spirits, of turpentine. A greater or less quantity than the above may be mixed up, in the same proportion, according to the number of sheep affected. Put the quicksilver and Venice turpentine into a mortar, or small pan, which beat together until not a particle of the quicksilver can be discerned; put in the oil, or spirits of turpentine, with the hogslard, and work them well together until made into an ointment. The parts of the sheep affected must be rubbed with a piece of this salve about the size of a nut, or rather less. When the whole flock is affected, the shepherd must be careful in noticing those that show any symptoms of the disorder, by looking back, and offering to bite, or scratch the spot; and if affected, he must immediately apply the ointment, as it is only by paying early and particular attention that a flock can be cured.

TO PREVENT THE SCAB.

Separating the wool, lay the before mentioned ointment in a strip, from the neck down the back to the rump; another strip down each shoulder, and one down each hip; it may not be unnecessary to put one along each side. Put very little of the ointment on, as too much of it may be attended with danger.

To Destroy Maggots in Sheep.

Mix with one quart of spring water, a table spoonful of the spirit of turpentine, and as much of the sublimate* powder as will lie upon a shilling. Shake them well together and cork it up in a bottle, with a quill through the cork, so that the liquid may come out of the bottle in small quantities at once. The bottle must always be well shaken when it is to be used. When the spot is observed where the maggots are do not disturb them, but pour a little of the mixture upon the spot, as much as will wet the wool and the maggots. In a few minutes after the liquid is applied the maggots will all seek to the top of the wool, and in a short time drop off dead. The sheep must, however, be inspected next day, and if any of the maggots remain undestroyed, shake them off, or touch them with a little more of the mixture.

A little train oil may be applied after the maggots are removed, as sometimes the skin will be hard by applying too much of the liquid. Besides, the fly is not so apt to strike when it finds the smell of the oil, which may prevent a second attack.

This method of destroying maggots I have always found superior to any other, and it prevents the animal from being disfigured by clipping off the wool, which is a common practice in some countries.

---

* A preparation of Mercury called 'corrosive sublimate.'

FINIS.

www.ingramcontent.com/pod-product-compliance
Lightning Source LLC
Chambersburg PA
CBHW062353220526
45472CB00008B/1791